GENES AND GENDER
SECOND IN A SERIES
ON HEREDITARIANISM AND WOMEN

Series Editors
Betty Rosoff and Ethel Tobach

Gordian Science Series Number II

GENES AND GENDER: II
PITFALLS IN RESEARCH
ON SEX AND GENDER

Edited by
Ruth Hubbard and Marian Lowe

GORDIAN PRESS
NEW YORK
1979

GORDIAN PRESS, INC.
85 Tompkins Street
Staten Island, N.Y. 10304

First Edition

**Copyright © 1979 by
Ethel Tobach and Betty Rosoff**
Series Editors

Library of Congress Cataloging in Publication Data (Revised)

Genes and Gender Conference, 1st, New York, N.Y., 1977.
    Genes and gender.

    (Gordian science series; no. 1)
    Vol. 2 has special title: Pitfalls in research on sex and gender, edited by R. Hubbard and M. Lowe.
    Includes bibliographies.
    1. Sex differences (Psychology). 2. Sex role. 3. Sex — Cause and determination. I. Tobach, Ethel, 1921-
II. Rosoff, Betty. III. Title. IV. Title: Pitfalls in research on sex and gender.
BF692.2.G46   1977   301.41   78-50640
ISBN 0-87752-219-7 (y. 1)

CONTENTS

PROLOGUE

The first volume of *Genes and Gender* dealt with the myth of genetic determinism as it was used to exploit women in our society. It gave the facts about known actions of genes in the development of bodily structures and function, about the way hormones work and the ways in which these facts are distorted by many biologists, psychologists, anthropologists and other scientists to "justify" the inferior social roles (gender) assigned to women, and the low value placed on their cultural, economic and political contributions. The authors in that first volume introduced information to refute the myth of genetic destiny flowing from Wilson's sociobiological theory (Wilson, 1975). It is fitting that the present volume of *Genes and Gender* develops this theme and presents an insightful critique of sociobiology, the "new" theory of genetic determinism.

Many biologists, economists, psychiatrists and other members of the scientific and professional community have uncritically adopted the old ideas put forth in "new" form by Wilson. Sociobiology perpetuates the assumption that people have much to learn about governing themselves by studying insects, apes and other animals. Sociobiological theory also argues that the roles female animals play in adjusting to their environment and in producing offspring is the "natural" model for women.

Magazines, movies, TV shows, and radio repeat this propaganda over and over. The media tell us that "scientists" have *proven* that women are limited in their abilities to do what men can do "naturally." The second volume of *Genes and Gender* shows how unscientific those "proofs" are. The authors point out that the research stands on quicksand—the quicksand of prejudice against women and minorities. The research on "sex differences" at the least wastes the time and effort of people who might be better employed in trying to understand the root causes of human problems. At the worst, it is an attempt to "prove" that "nature" programs women to be inferior to men through gene action.

Women have to expose the false "scientific" research, and the articles in *Genes and Gender II* will give them the material to do so. We welcome the contributions of Ruth Hubbard, Marian Lowe and their colleagues who have worked so hard and so well to carry the mission of the first *Genes and Gender* Conference forward.

Ethel Tobach
Betty Rosoff
September 26, 1978

INTRODUCTION

The chief distinction in the intellectual powers of the two sexes is shown by man's attaining to a higher eminence, in whatever he takes up than can woman—whether requiring deep thought, reason, or imagination, or merely the use of the senses and hands. If two lists were made of the most eminent men and women in poetry, painting, sculpture, music ... history, science, philosophy, with half-a-dozen names under each subject, the two lists would not bear comparison. We may also infer ... that if men are capable of a decided pre-eminence over women in many subjects, the average of mental power in man must be above that of woman ....

Charles Darwin, 1871

Women are never supposed to have any occupation of sufficient importance *not* to be interrupted except "suckling their fools;" and women themselves have accepted this ... and have trained themselves so as to consider whatever they do as *not* of such value to the world or to others, but that they can throw it up at the first "claim of social life." They have accustomed themselves to consider intellectual occupation as merely selfish amusement, which it is their "duty" to give up for every trifler more selfish than themselves ... If a man were to follow up his profession or occupation at odd times, how would he do it? It is acknowledged by women themselves that they are inferior in every occupation to men. Is it wonderful? *They do everything* at "odd times."

Florence Nightingale, 1859

In recent years there has been a veritable deluge of pronouncements purporting to demonstrate scientifically the biological basis of various behavioral differences between women and men. Many of the authors explain the unequal positions occupied by women and men in our present society as being due to these inherent biological differences. The papers in this book are part of a growing body of work raising fundamental questions about the scientific as well as the political content of much of the contemporary research into sex differences.

Our main concern, however, is less with conclusions than with methodologies. Our primary aim is to try to determine whether it is possible to design experiments capable of yielding decisive answers regarding the contributions that genes make to gender, given that we live in a society in which from the moment of birth a child's sex is regarded as its most important characteristic — indeed, in which any trace of sexual ambiguity is regarded with horror. Since, however, the theories are inseparable from their social context, in addition to examining various aspects of this question, the papers also deal to some extent with the social implications of the conclusions.

Attempts to define the origins of sex differences in behavior are not new. Throughout documented history, and probably before that, people have tried to understand why we behave the way we do and what governs our destinies in the long run. In pre-scientific days gods, stars, and various juxtapositions of real or imagined phenomena were looked to for explanations. Since the rise of modern science these kinds of explanations have lost favor, though some of them were empirical and based on repeated coincidences and observed correlations. In recent times, since the rise of individualism and liberalism, people in the West have tended to seek explanations for their actions within themselves and have looked to biology and psychology to provide them. Indeed, they have wholly, or in part, created these fields to provide such answers. Looking inward, while at the same time searching for unique causes, scientists have turned to our genetic make-up, our prenatal or birth experience, our first maternal contact, or even the hormone concentration in placental blood as a possible explanation of our behavior years and decades later.

Interest in the reasons for human behavior and human differences has risen and fallen during different periods as social conditions have changed. As one looks at the history of the attempts to explain the origins of sex differences, it is clear that in previous times, just as at present, interest runs particularly high during times when women's social, economic, and political roles are in rapid transition.

Why are differences interesting?

When we try to understand the reason for the present wave of interest in the biological basis of sex differences, it is important to realize that as long ago as about 200 years several important changes that reinforced one another occurred in Western Europe and in its American colonies. The bourgeois revolutions on the two continents proclaimed the Rights of *Man*, and Liberty, Equality and *Fraternity*. These were no generic uses of words. The *Fathers* of the revolutions had no intention of sharing these rights and freedoms with their wives, daughters, and sisters. Yet, irrespective of their intentions, the proclamations and the political changes that accompanied them spurred women to greater efforts in their own behalf. Mary Wollstonecraft wrote her feminist manifesto, *A Vindication of the Rights of Woman*, as an outraged response to Rousseau's prescriptions for the differential education of boys and girls as he described them in *Emile*; Abigail Adams tried to get the Constitutional Convention to "remember the ladies," albeit through her husband, John, who did not take well to her counsel. By the middle of the nineteenth century, as industrialization began to bring major changes into women's lives, the feminist movement was rolling along. Nor could women's arguments for equality be countered easily by the traditional strictures of church and state, for both were losing that kind of authority.

Throughout the nineteenth century God's Laws were gradually replaced by Laws of Nature as science increasingly gained authority to define reality — an authority that resulted in no small measure from the economic successes reaped by the ascending merchant class as a result of the applications of science and technology. However, these very successes involved economic transformations that moved the locus of production from the home to the factory and in the process accentuated and rigidified a sexual division of labor as well as class differences. At the same time the new outlook on reality emphasized individual qualities rather than social forces as determinants of a person's actions and her or his successes or failures. The juxtaposition of this liberal ideology of the upper classes with the harsh social realities of nineteenth century industrialism and colonial exploitation created an urgent need to demonstrate a legitimate basis of the observed differences in status and power between women and men, poor and rich, subject and colonist (or in the United States, non-white and white). Scientists in particular increasingly felt called upon to explain these differences.

It is important to recognize that science and social ideology exist in a complex interaction, which we discuss more fully later. For now, it is enough to point out that scientists of the nineteenth century were not hypocrites who were willing to distort their science to prove a political point. They were almost entirely university trained, privileged sons of the upper and middle classes who accepted the social order within which they flourished as founded on natural law that science could be called on to explicate. Science was not merely a body of information, but a whole ideological framework in which the world could be explained. By the late nineteenth century scientific explanations, values, and methods of inquiry dominated every area of social thought. The successes of the physical sciences strongly influenced the new fields which appeared during the century: anthropology, sociology, and psychology, among others. The study of society became the "science of society" or "social science" and the theoretical framework of interpretation of the social world was based on physical or biological analogies. The goal of these new sciences was to provide answers to major social questions.

Scientists were interested in applying the tools at their disposal to investigate the laws of the social order — particularly when the disaffected clamored for change. Their belief that their society was founded on biological or physical laws made it reasonable to look for the intrinsic biological differences that would explain why groups fared so differently in the world. In doing so they did not acknowledge that the ways in which observations are made and, indeed what one sees, are strongly colored by the hypotheses one uses and by the framework in which one's observations are made. If one is sure that there are innate behavioral differences and tries to demonstrate this, then facts which do not support this belief are often simply not seen or are even unconsciously distorted or misinterpreted.

What did science say about sex differences in the nineteenth century?

Early nineteenth-century theories show clearly the kinds of assumptions upon which scientific reasoning about sex differences has been based, both then and later. Women were seen as fundamentally different from men in a multitude of ways. A woman had a smaller cranium, weaker muscles, smaller bones, and more delicate tissues. The female nervous system was viewed as very irritable due to its complex and careful organization. A woman's most important biological characteristic, however, was her reproductive system and all aspects of her biology were seen ultimately as related to reproductive functions. These attitudes, present through much of the century, led a later prominent gynecologist to comment that it was "as if the Almighty, in creating the female sex, *had taken the uterus and built up a woman around it.*"[1]

The claim, accepted as scientific fact, that women's nervous system was especially sensitive, was based on no direct scientific data, but came from observations of social behavior such as heightened sensibilities and "nervousness." Once the existence of a physical difference in the nervous system was accepted, it was then used to "explain" the same social behavior. Such ideas, furthermore, were used not only to explain but also to prescribe proper female behavior. The assertion that the female nervous system was particularly irritable was translated into a picture of woman as emotional, fragile and dependent; the theory that a woman had an intricate and centrally important reproductive system meant that *all* her energy and effort had to go into it if its enormous energy needs were to be met. She had to direct herself first and foremost toward motherhood and not waste her forces or else the race would die out or decline in vigor.

While theories about women continued to emphasize the female reproductive role as the chief source of female and male behavioral differences, later in the century the kinds of arguments and evidence changed. Most prominent after mid-century in explaining human behavior, and sex differences in particular, were evolutionary biosocial theories. Social Darwinism, an attempt to understand human societies through adaptations of theories of evolution by natural selection, provided a unifying framework for much of the work that was done in the new social sciences. The underlying assumption in most of the work was that the society of the period was the culmination of an evolutionary process. Different societies arose because of the different characteristics of the people that lived in them, and one could assign them a place in an evolutionary hierarchy, some being in a more advanced stage of evolution than others. Once this assumption was made, work focused on finding out why a particular behavior had evolved or on determining the biological, evolutionary basis for observed differences between societies or individuals. There was no room in these theories for the idea that perhaps a given behavior did not result from the evolution of a physical or biological characteristic, but had a social cause. Therefore, these evolutionary theories inevitably provided an argument against social reforms, such as equality for women. Since societies, like organisms, were shaped by biological evolution, the idea followed that the proper course of evolution should not be interfered with. It was argued that any attempt to do so would probably result in lessened biological fitness, both individual and societal. (But it is important to understand that the scientists who argued this were not disinterested outsiders. They were white, upper class men and therefore stood to lose from changes that might have equalized the distribution of power and privilege in the society.)

Evolutionary theories all attributed sex differences to different evolutionary pressures on the sexes.[2] It was claimed that woman's distinct emotional nature was built around her "maternal instinct" and was a direct consequence of her reproductive physiology through the operation of sexual selection. Darwin held that such human sex differences arose through natural selection with men being adapted as hunters and protectors and women as mothers. Spencer added that woman's physical evolution had stopped at an earlier stage than man's in order to conserve her energies for reproduction. At the same time, women had made psychological adaptations during the struggle for existence which had given the women who survived the ability to survive the aggressive and egotistic tendencies of strong, brutal men. Like Darwin, Spencer saw women as somewhat childlike in both mind and body.

Later, a widely publicized and influential theory by two biologists, Patrick Geddes and J. Arthur Thomson was more specific about the way in which the evolutionary mechanism operated. Using information from a wide variety of organisms, they derived sex differences from different physiological activities of the sexes. For example, males were characterized by a tendency to dissipate energy, females by the ability to store energy. Despite inconsistencies in their data which they had trouble interpreting, they claimed to have shown that all sex differences depend on the greater activity of the male organism, and that this was true even of the lowest forms of life. The sex differences had existed from the beginning, and fundamental change was not possible. "The differences may be exaggerated or lessened, but to obliterate them it would be necessary to have all the evolution over again on a new basis. What was decided among the prehistoric Protozoa cannot be annulled by Act of Parliament."[3]

Other scientists searched for some solid evidence of the differences that evolutionary history was assumed to have created. Work in anthropology is representative both of the science of the period and of the motivation behind it. The problems of demonstrating racial and sexual differences occupied much of the attention of nineteenth- and early twentieth-century anthropologists. Much of the early work was done on race, but in 1869, anthropologists, fearing that demands for female rights threatened to divert the orderly progress of evolution, decided to turn more attention to the position of women and to its origins.[4] The anthropologists believed that social reformers' ideas about the intellectual equality of women violated nature and would have a "most baneful effect in unsettling society." Women were clearly inferior intellectually, and it was now time to protect them by using science, rational and objective, to demonstrate once and for all women's true place in nature, so that they would be able to live in accordance with their biological destiny, which was of course motherhood. The social reformers were seen as anti-scientific and misinformed, as indulging in metaphysics. (A second motive for considering the "woman question" was a wish to make anthropology relevant to the contemporary social and political issues.)

A promising tool for this investigation seemed to be skull measurement, or craniology, which had already shown considerable "success" in demonstrating racial differences. S. J. Gould recently has re-analyzed data gathered by Morton, the most prominent of those working on racial ranking. He suggests that Morton's conclusions were determined by "an *a priori* conviction of racial ranking so powerful that it directed [his] tabulations along pre-established lines,"although the data did not warrant his conclusions.[5] The conviction of female inferiority was similarly powerful in directing the work of anthropologists when they sought to use craniology to study sex differences in intelligence. (The forerunner of craniology, phrenology, had already been widely used to demonstrate sex differences.)

Similarities between measurements for women and for so-called primitive races were quickly seen by the distinguished French scientist and surgeon, Paul Broca, among others. Dimensions for both groups were found to be less than those for white males and this was taken to be evidence that both were less intelligent and on a lower evolutionary level. (The deficiencies of the female and non-white brains were thought, not only to make them less intelligent, but to allow more "primitive" aspects of "human nature" to appear.) Unlike the case of race differences, however, a sex difference in absolute size of both skull and brain could be demonstrated without manipulation of the data. If intelligence depended on brain size, an assumption widely accepted when discussing racial differences, then women were clearly less intelligent than men. However, since males were both taller and heavier on the average than females, the "elephant problem" surfaced. If absolute size of the cranium were taken as the criterion, then elephants and whales ought to be more intelligent than humans. The most obvious solution to this problem was to use brain weight relative to body weight as a measure of intelligence. However, this proved to be an unsatisfactory measure. The work of Christian Bischoff, for example, suggested that women's brain to body weight measurement was 6% greater than men's. Once the superiority of women was suggested by the brain to body weight studies, scientists moved quickly to try other criteria which might demonstrate the "correct" relationship of the sexes. A great assortment of indices were tried, including brain weight to body height, the ratio of brain to thigh bone weight, and the ratio of brain to spinal marrow, but these were increasingly unsatisfactory.[6] Even on those indices where men came out ahead, the relationship to intelligence was hardly obvious.

At the same time that measurements of the size of the whole brain were taking place, other scientists were working with specific areas of the brain, attempting to show that one lobe or another was different in size in females and males, and connecting their results with difference in intelligence or emotion. This work ran into difficulties similar to those in the work on the whole brain and it proved to be equally unsatisfactory.

By the beginning of the twentieth century craniology finally was moved out of the mainstream of science. During the period of its dominance, it appeared as the embodiment of rationality and objectivity. Ultimately it foundered and was abandoned, not because it was seen as unscientific, but because it appeared to be an unprofitable approach to finding an index that could be tied to differences in average intelligence between groups, one which would consistently put white men ahead — the place where anyone looking around could see they belonged. Eventually the search for physical indications of intelligence differences was abandoned and psychologists began to develop "mental" tests, such as the IQ tests.

After the first decades of the twentieth century, as measurement or observation failed to confirm their hypotheses, most of the biosocial evolutionary theories no longer occupied influential positions. However, a conviction that natural behavioral differences existed between women and men still shaped the behavioral sciences. While the earlier theories were losing their appeal, the new fields of reproductive physiology and sex endocrinology emerged and grew tremendously.[7] Focusing on research on the newly discovered sex hormones, these fields offered a brief flicker of hope that sex differences could once and for all be reduced to differences in a critical molecule, *the* male or female hormone, *androgen* or *estrogen*. Biologists automatically assumed that the formation and functions of these hormones were sex-specific, that one was made in the male, the other in the female, and that each governed the development of the sex appropriate characteristics. The extension of the model to sex-specific social roles was obvious.

The theory of sexual polarity (or dichotomy) which was built into the ideas and research in biology and endocrinology, for a time went virtually unquestioned. However, the hope of such a simple solution soon was dashed by the discovery that both hormones occurred in both sexes, and more recently by finding that body tissues can convert one into the other, so that one cannot be sure which is active in a given place or at a given time.

There is a large body of work in the nineteenth and early twentieth centuries on sex differences which we are now beginning to learn about through historical studies stimulated by the current women's movement. Like the examples given here, all the work, in one way or another, ascribed sex differences in behavior to the needs of women's reproductive function. This preoccupation with the reproductive system led to the idea that all female ailments, physical and psychological, could be traced to its malfunctioning. The medical focus on women's reproductive system, in turn, reinforced the social emphasis on the centrality of women's reproductive role. Medical technology, increasingly influenced by the rise of scientific thought, was then used to try to cure these difficulties. Psychological problems, particularly deviations from proper feminine behavior, as well as physical complaints were treated by drastic medical and surgical techniques that concentrated on women's "delicate" reproductive systems.[8] Medical treatment included an assortment of washes, injections and such procedures as the application of leeches and cauterization with hot rods or chemicals. Increasingly common were surgical procedures: the ovariotomy (removal of the ovary), clitoridectomy (removal of the clitoris) and hysterectomy (removal of the uterus). The operations were often performed on healthy organs in order to "cure" psychological problems. These drastic treatments served to reinforce ideas about women's nature and role. They also introduced the idea, still with us, that socially created behavior problems, defined as illnesses, could be cured by medical technology.

After the triumph of conservatism at the end of the "Progressive Era," the first wave of the women's movement ebbed and with it interest in the origins of sex differences on the part of the more influential biological and behavioral scientists. By the 1930's, the emphasis of scientific and psychological questions about women had shifted to the measurement of motivational and personality traits. These developments paralleled the increasing acceptance of Freudian psychology with its emphasis on the different personalities, motivations and emotional needs of women and men. Since Freud had attributed these differences primarily to biological factors, the question of inherent differences between women and men was not totally neglected, but most studies did not consider it explicitly.

During the 1950's and early 1960's, as the Feminine Mystique became dominant, experimentalists shifted to measurements of developmental processes involved in the formation of sex roles. The question of whether sex differences in behavior were inherent or environmental seems to have been considered an open one, interesting but difficult.

Why the renewed interest in sex differences?

In the last decade, parallel with the new flowering of the women's movement, there has also been a renaissance in research on sex differences and on the roles of nature and nurture in determining women's and men's social roles. For, once again, a wide-ranging interest in the origins of our social behavior and social institutions has appeared along with the many challenges to the existing social structure — those threats to social stability evidenced not only by the women's movement, but by the demands for equality on the part of Blacks and other minorities, by the anti-war movement, and by the multiple challenges to the American Dream that have grown out of the recurring (or rather, continuous) economic crises.

The increasing attention that has been directed at sex role behavior, however, probably is due only in part to women's louder demands for equality. In part it must be attributed to the increasing realization of the profound economic implications of having large numbers of women, particularly those with small children, enter the work force. A recent front page article in *The New York Times* was headlined, "Vast Changes in Society Traced to the Rise of Working Women." In it, Eli Ginzberg, chairperson of the National Commission for Manpower (!) Policy was quoted as stating that the "revolution in the roles of women . . . will have an even greater impact than the rise of Communism and the development of nuclear energy. It is the single most outstanding phenomenon of this century."[9]

In spite of the realities of women's large numerical participation in the work force, now, as in the nineteenth century, scientists are trying to find intrinsic biological "reasons" why men go to work and women stay (or *should* stay) home. And interestingly enough, the search now centers in research areas that are direct descendents of the earlier work which we have mentioned, areas that are furthermore maximally fraught with ambiguity, uncertainty and opacity. One set of theories (sociobiology) attributes the differences in roles to differences in the evolutionary demands on the two sexes owing to differences in courtship rituals and in reproductive strategies and functions; another group of theories deals with purported differences in brain structure and functioning; a third group tries to identify *critical*, differential hormonal effects. Of course these searches for causes do not develop in isolation, so that combinations of them are also found: differential effects of hormones on the brain; differential evolutionary effects on hormones; and so on. The important thing to realize is that all three areas — evolution of behavior, brain-behavior interactions, and hormone-behavior interactions — pose at present (and perhaps even permanently) insurmountable difficulties for the design and implementation of clean and decisive experiments. Yet, not surprisingly, even the most tentative results in these highly ambiguous (but charged) areas are hailed as major breakthroughs in the press, particularly when they reinforce the status quo.

In the current revival of nineteenth century theories, the postulate of some ethologists that sex-differentiated social institutions have evolved in response to the evolution of biologically based behavioral differences between females and males were the first to gain widespread attention. Many modern ethologists assume that the evolutionary history of a species determines its behavioral "potential" and they attempt to study this relationship. They argue that since humans have evolved from apes and other "lower" animals, important information about our behavior can be obtained by studying theirs. Groups of animals, called "societies," in which behavior is assumed to be innate (an increasingly problematic assumption) are used as models of human society. (Of course, even the term, human society, is a reified, ideological construct. There is no such thing as *a* human society; there are only different human societies that are the historical outcomes of the ways in which different groups of people have worked and lived.) The presumed coincidence between animal and human "societies" is then assumed to prove that biology dictates the behavior of people as well as of the animal models. Several popularly written books that presented these kinds of arguments during the 1960's became or just missed being bestsellers. Examples are *African Genesis, The Territorial Imperative* or *The Social Contract* by Robert Ardrey; *The Naked Ape* and *The Human Zoo* by Desmond Morris; *On Aggression* by Konrad Lorenz; *Men in Groups* by Lionel Tiger; or *The Imperial Animal* by Lionel Tiger and Robin Fox. Their message is that human competition, territorial greed and aggression to the point of war, monogamy and women's submission and dependence on men are natural, biologically based and unavoidable. No wonder people wanted to read them in hopes of finding biological answers to the doubts and questions raised by the Black and Women's Rights Movements and by the Vietnam War!

Although animal ethology has gained wide scientific recognition (the 1974 Nobel Prize for Physiology or Medicine was shared by three ethologists, including Konrad Lorenz), the application of ethology to human societies has been criticized by many biologists as simplistic and "unscientific." As a more scientific improvement, in 1975 Harvard biologist E. O. Wilson put forth a new version of sociobiology. Like ethologists, sociobiologists try to understand the biological basis of animal and human behavior by using an evolutionary model. However, they claim to have gone beyond the "advocacy" theories of works such as those mentioned above and to have put the field on a firm scientific footing. This claim is examined and criticized in the papers that follow by Bleier, by Leibowitz, by Salzman and by ourselves.

Like evolutionary theories, the study of brain structure and function to explain sex, race and class differences has also come in for a revival. In 1969, the psychologist, A. R. Jensen, announced in a widely publicized article that differences in achievement between Blacks and Whites derived from innate measurable differences in IQ between the two groups, a claim first made when IQ tests were developed in the first decades of this century but that had been dormant for fifty years. Jensen's revival was taken up by William Schockley (Nobel Laureate in physics for his development of transistors) and in Britain by the psychologist, Hermann Eysenck. At Harvard, Richard Herrnstein extended the generalization to class and sex differences. Much of the subsequent controversy centered around the claim of racial differences, for IQ turns out not to be as "useful" as a measure of behavioral differences between women and men. A single number cannot be assigned to sex differences as it can to race differences. Rather, the relative performances of females and males differ on various parts of the standard IQ test, with females ahead on some of them and males on others. Indeed, an embarrassing and little known fact is that when IQ tests were originally set up, although the class and race differences were in the "right" order (that is, upper class whites ahead), girls consistently scored better than boys. The tests were, therefore reweighted so that now there is no sex difference overall, while race and class differences remain. (A question: Why was this equalization of female and male scores deemed appropriate and scientifically valid, whereas the purported difference between Blacks and Whites has been heralded as the "explanation" for Blacks' inferior economic and social status?)

Lest we leave our readers with a misunderstanding, it should be stressed that differences in IQ scores are arbitrary. No one contests that one can construct "IQ tests" on which Blacks score higher than Whites, or the reverse; and so on. The IQ test is a peculiar instrument that measures unspecified parameters in the cognitive space of the person being tested. It can be structured to favor one or another way of learning and understanding, and tells one nothing of interest about "intelligence" or about any other significant biological or psychological trait. Current tests were constructed to favor those traits correlated with success in our schools. They are political tools, not scientific ones. Mental testing of this kind, however, has not proved a very successful tool in examining sex differences, and we therefore are now seeing a revival of work on physical indices of cognitive abilities. Much of the interest in brain differences between women and men now concerns measurements of hemispheric dominance or brain lateralization — the ways in which the right and left halves of the human brain interact. Like the nineteenth century explanations that were based on brain structure, some of the present explanations try to derive observed sex differences in behavior and abilities from presumed intrinsic differences in brain structure and function. In her article, Star describes some of the new work and discusses its methodological problems and political implications.

The third important area in the ongoing work that tries to identify the biological basis of sex differences in behavior involves the so-called sex hormones. Given the known complexities of the changes in hormone levels and interactions, a more sophisticated theory than the early twentieth century one is required. Since deliberate experimentation with human hormone levels is impossible, the actual experiments must be done with laboratory animals whose hormone levels can be grossly manipulated, and the results extrapolated to humans — an extrapolation which in general is impermissible. Observations also are made with children who, by virtue of pathology or unfortunate medical intervention, have been exposed *in utero* to abnormally high or low concentrations of one or another "sex hormone," so that they are born with such severely malformed external genitals that their sex is ambiguous or even misassigned. By observing such children, John Money and his colleagues claim to have shown that sex hormones "program" the brain differently in girls and boys. They further suggest that this difference may be responsible for a number of the specific behavioral differences that are seen in our society. These lines of work and theorizing are analyzed in the articles by Bleier, by Salzman and by ourselves.

An example of an evolutionary-hormonal hybrid theory is that of the sociologist Alice Rossi, who has combined an evolutionary theory similar to E. O. Wilson's with the assumption of sex-differentiated hormonal programming of the human brain to update the nineteenth century concept of the "maternal instinct." The problems with her work, which she presents as a new, interactive, biosocial theory, are discussed in our paper.

Raymond, in her paper, discusses one extreme consequence of these kinds of efforts to root behavior in biology: the creation of the "disease" syndrome, transsexualism, and of its medical "cure" through hormone treatments and sex change operations. Like ovariotomy and clitoridectomy, this is an overt effort to resolve psychosexual and social conflicts and suffering by restructuring the bodies of the afflicted individuals.

How does this kind of theorizing come about?

Why have scientists so often come up with theories of innate sex differences in behavior? Is it because that is simply the way it is? As the papers in this volume demonstrate, not only can the question, whether behavioral sex differences are due to nature or nurture, not be answered, but it is a pseudo-question; it has the grammatical construction of a question, but it is based on false assumptions and hence is scientifically meaningless. It is not the right way to ask the question, and theories which attempt to answer it in this form are not scientific theories, but propaganda in the guise of science. We discuss this point in our paper, as does Salzman in hers.

If the work is flawed as science, but continues to be done and repeatedly is used to demonstrate innate behavioral differences, the existence of which can be made to explain and legitimize the existing social order, is this because the scientists who do it are sexists (or racists) and consciously try to bolster the dominant ideology? Some of them may in fact be, but as we have already suggested when discussing the sexist and racist theories propounded in the last century, there is a more fundamental problem which stacks the deck in favor of this kind of biological determinism. To understand this problem better, we must look more thoroughly at what science is, how it is done, and at the interactions of science and scientists with their society.

In the past few centuries, as science has increasingly been accepted in the West not only as one among several descriptions of the natural world, but as the only correct one, many people — including scientists — have lost their awareness of the fact that science is a human construct. Science has been reified (that is, literally, turned into a thing), as though it existed out there in the world and had a reality of its own; indeed as though it were not a way of describing natural phenomena, an abstraction created by the human mind, but as though it were nature itself. However, the fact is that any observation or description of natural phenomena involves a process of selection. Nature is a continuum that includes us humans within its interdigitating web. To take ourselves out of the picture in order to try to observe what goes on around us already involves a distortion of reality. What objects and activities we notice or leave unnoted involves further selections, albeit usually largely unconscious ones. Making the selections requires a conceptual framework — reasons why this fact is important and that one is not. What, among all the things and events on which we focus our attention so that we "observe" them, we choose to make the "data" which we then integrate into the coherent system of explanations that constitutes modern science requires yet a further set of choices. These choices again are based on preexisting theories of what is real and important.

In other words, science is a product of the human imagination, created from theory-laden facts. Furthermore, science is not created anew by every generation of scientists. It is based on, and linked with, the science that has gone before and is the foundation of the science to come. Its building blocks are abstractions, or rather concretizations, reifications and isolations from nature which are knit together by the scientist's effort to build a self-consistent system of explanations. It is clear that this system must have cross-linkages to the real world out there, because it "works." The laws of physics allow us to build machines that do predictable things, and this gives us some assurance that though physics may be only a partial and selective description of the world, it is one that contains important correspondences to it. But such reality tests are much more inadequate and often entirely lacking in biology, particularly the part of it that is concerned with behavior; in the social sciences the situation as far as reality tests are concerned is even worse. (Indeed this is one reason why some people worry about new biological technologies such as recombinant DNA. They are reality tests; and what if the presumed correspondences are wrong?)

Science is, as Alan Watts has said, "an awareness of nature based upon the selective, analytic, and abstractive ways of focussing attention. It understands the world by reducing it as minutely as possible to intelligible things."[10] The important point for our present discussion is that to be "intelligible," the "things" have to fit into the general social, political and philosophic context within which the science-making person — the scientist — lives. One must have a framework in order to make decisions about what to notice and in order for the things one notices to be intelligible. But this framework itself can shape not only what is observed, but also how it is observed. In *The Structure of Scientific Revolutions*, Thomas Kuhn has pointed out the importance in the physical sciences of the "paradigm," the set of assumptions, theories and laws that form the framework of a particular field of science.[11] Once the world is ordered and made intelligible by this paradigm, facts which do not fit into it have a hard time intruding themselves on the consciousness of the observer. In order to preserve the paradigm, facts which in reality contradict it may be seen by an observer as actually supporting the system of beliefs. Kuhn gives an example in which observers were asked to identify playing cards flashed before them, some of which, unknown to them, were not the conventional ones. For example, one might be a *red* ace of spades instead of the conventional black one. Yet the observers consistently fitted the anomalous cards into their "paradigm" of what playing cards are like: they actually *saw* a black ace of spades, not a red one. They were not able to see a fact that contradicted what they *knew* to be true.

The effect of belief in a paradigm becomes particularly important and decisive when it comes to making biology or one of the social sciences, the disciplines that try to discern, describe, catalog and analyze the activities of other living organisms and of people. Scientists' social perspective and experiences inevitably cloud their sight and guide what they deem worthy of notice, as well as the meaning and validity of what they notice.

Anthropologists, who are trained to be aware of the hazards of internalized socialization for the keenness of one's perceptions, use the terms, foregrounding and backgrounding, to denote the often unconscious activity of allowing certain things, habits and occurrences to fade into the background of the unnoticed or "unnoteworthy," while pulling others into the foreground of one's analytic canvas where they stand out enough to command attention. And anthropologists also know how soon upon entering an unfamiliar culture the pattern of selectivity within which one notices things changes, as well as the interpretation of what is noticed (which is not to say that what one sees need be any more "true" or "real" initially than at the end).

When a child is repeatedly admonished to make such a subtle distinction as "this is a boy, not a girl," the very subtlety, coupled with the insistence that the distinction be made correctly and every time, conveys a profound social message. The distinction also provides a fundamental framework from which to view the world. In another culture or in another natural setting, this kind of emphasis may be put on different textures of snow or on the different sounds made by water. It just depends on what is important in one's life and in one's particular setting; but the result will be worlds that are conceptually and experientially very different.

We often internalize the most profound and significant concepts, then, quite without knowing it. Usually only people who are, for some reason, marginal to the prevailing belief system — radicals of some sort, or women who have become aware of their historic exclusion from the process of reality-making — are even tempted to become conscious of the process and to question its results. The attempt to bring internalized assumptions to awareness, of course, is what "consciousness raising" is about.

Let us look at the conceptual framework within which behavioral science is carried out in Western culture. Group identity is an important part of our social reality. A person's position in our highly structured, hierarchical society is strongly affected by sex, race and ethnic background. People are educated to think in terms of categories and stereotypes, and for most public purposes, sex and race have become more important than almost any personal, individual characteristics. In this, scientists are no different from anyone else, so that for behavioral scientists from our culture looking at human or animal societies — even those that are structured very differently from ours — group division and hierarchy definitely are part of the foreground. Furthermore these divisions are perceived from within a social mind-set that makes them seem inevitable, normal and, at least in part, desirable.

The tendency in our society to identify individuals at once as group members — women, Blacks, Irish, Jews — is reinforced for scientists by the strictures of the scientific method. For one of the chief distortions that the scientific view introduces into reality derives from scientists' need to analyze, discriminate and categorize. Through their training, scientists learn to define and describe differences better than commonalities and to find them more "interesting." Indeed a finding of "no difference" often is considered so insignificant that it is not worth mentioning. Therefore, in the effort to find something of significance, behavioral scientists tend to continue to look until some difference turns up that is worth describing. And of course, if one looks for differences between two groups that occupy different positions in society, one is almost certain to find some.

But although it is not difficult to *find* differences if one looks hard enough, there is no theoretical framework within which to evaluate their intrinsic significance for social interactions and social differentiation. In the absence of such a framework, *any* sex difference can be asserted to be significant or even decisive for a particular form of sex-differentiated social behavior.

Moreover, a common mistake of attempts to build such a theoretical framework is to treat the behavior as though it could be separated from its social context. For outside this context, the observations are nothing more than celebrations of what C. Wright Mills has aptly called "the abstracted almighty unimportant fact."[12] For example, it is asserted that there is a sex difference in dilation of the pupils of the eyes when people are shown pictures of infants and that this is evidence of a different *biological* response to infants on the part of females and males, as though social context did not matter. However, without some idea of sex differences in the expectations about parenting behavior, this observation is meaningless. Within the context of our society, on the other hand, the observation (if indeed it is correct) must be seen as a correlate of differential behavior of actual or potential mothers and fathers; and in this context, its origins are no clearer than are those of the parenting behavior to which it is assumed to be related.

This also illustrates a more general methodological problem: the reification of a "trait" — parenting, aggression, dominance, maternalism, etc. — as though it were an object or a category of objects. The next step then is to look about and find examples. But in the absence of adequate theoretical constraints, one's examples easily become a hodgepodge of *ad hoc* conjunctions that are united only by the fact that one has put them together. This is how a concept like "male aggression" is substantiated by looking at the grooming of baboons, the trooping of the colors, and social displays of Rocky Mountain Bighorn sheep.

What can one validly say about sex differences?

It follows from what we have said that it is a central methodological question whether *any* valid results can come of work that sets about deliberately to look for sex differences. Hugh Fairweather ends his recent comprehensive review of research on sex differences in cognition with the following paragraph:[13]

> It must be stressed, finally, that the majority of studies reviewed here and elsewhere are both ill-thought and ill-performed. Whilst in other circumstances this may be regarded as the occupational hazard of the scientific enterprise, here such complacency is compounded by the social loadings placed upon these kinds of results. It is clearly very easy to include sex as a bonus factor in experiments which have little scientific merit. We cannot pretend that we are testing a theory of sex differences, since at present none can exist. *Legitimate studies of sex differences can only grow first out of observations of clear individual differences* in the investigation of salient psychological processes; *and second, from the observation that the groups of individuals thus differen-*

tiated have clearly biased compositions when divided by sex. Studies with clinical populations would already tend to meet these criteria. Studies within the normal population, predicated on the assumption that discriminations are useful, can only be regarded as tempting sexism. (our italics)

This stark statment applies to all studies of sex *differences*, because it is methodologically hazardous to try to prove the existence of differences by looking for them. A research design that is so selective, is all too likely to become a self-fulfilling prophecy.

The fact is, of course, that there is a great deal of overlap between the sexes as regards both physical and behavioral characteristics. Except for anatomical and physiological differences directly connected with reproduction, women and men are a great deal more alike than they are different. Even for such an "obvious" sex difference as height, the average difference between women and men is a matter of a few inches (and very dependent on populations) whereas the range of heights for both sexes (within the norm) is over 3 feet. Obviously sex is not a very good predictor of height for any given individual. It is certainly not a good predictor of behavior. Here *average* female-male differences, where they exist, are small and have proved elusive and particularly difficult to measure. Yet, despite the almost complete overlap between the sexes on most traits that are not directly linked to reproduction, most behavioral scientists (like most other people) believe that behavioral sex differences exist, and tend to give them more emphasis and importance than they do to similarities. As L. E. Tyler points out in her book on differential psychology, "The research worker in human differences does not start out in a vacuum. His [or her] task is often not simply to make a beginning in a field where nothing is known, but to check up on convictions that are held with dogmatic certainty."[14] Behavioral scientists, like many other people, seem to take it for granted that where such blatant differences in social, economic and political power exist as those between women and men in the industrial, capitalist societies of the West, there must be intrapersonal — or rather, intragroup — reasons. And they see it as their job to find them. We have already discussed at some length the effects that such a conviction can have upon the way science is done, but there are some difficulties specific to the study of behavior and of differences in behavior in particular.

We have argued that one cannot legitimately abstract behavior from its social context. What remains to the scientist is to try to stitch together a "patchwork quilt" by abstracting enough of the context along with the behavior to make it meaningful. This involves choices which, as we have seen, are determined as much by ideology as they are by the scientific method. Similar difficulties beset attempts to determine the basic reasons why women and men behave differently in society by observing their responses to artificial tasks administered in artificial settings. By the earliest age at which humans will respond to external stimuli, they have already learned to do so in a social context that rewards socially appropriate or inappropriate behaviors differently. All "controlled" settings for studying behavior involve experimenters interacting in some way, direct or indirect, with those being studied. And in a society that has a host of emphatic preconceptions about how females and males *should* behave at all times, sex appropriate, albeit sometimes very subtle, messages get into just about all human interactions. This means that although laboratory settings may be artificial, they in general are *not* removed from the social context. (Indeed, if they were entirely removed, then the results would be meaningless for the study of human behavior.)

There is another way in which social values affect the science of differences: when a difference is established between groups that have different positions in the social hierarchy, the attributes of the dominant group are the "right ones" to have. For this purpose, any difference at all will do, even if it is only that the dominant group is blond and blue-eyed. In that case, being dark and swarthy becomes undesirable, and these "bad" characteristics are soon shown to be associated with other, well-known "bad" characteristics such as, for example, a lower IQ score. (Konrad Lorenz himself wrote in this vein when the time and place were right for that kind of "science").[15]

A good example of this point that is relevant to sex differences is a study in which a sex role stereotype questionnaire was given to a group of psychologists, psychiatrists and social workers. They were asked to identify characteristics of a normal, psychologically healthy adult of unspecified sex, of a normal man and of a normal woman. In their responses, they specified the same traits when describing a healthy adult or a healthy male; but the traits of a psychologically healthy adult and of a healthy woman were significantly different. Indeed the mature, healthy woman was assigned traits that the same observers rated as less healthy *human* traits than those of the mature, healthy male. In other words, the attributes of the dominant group (in this case, men) became the standard of healthy human behavior against which women were deemed inferior. Or, to say it another way, these professionals did not take account of the traits they thought of as normal women's traits when they established their "human" norms; hence, women's traits were judged by a male standard and were deemed *inferior* to it.[16]

What is the social role of sex difference research?

In thinking about sex difference research, and in reading the critiques in this book, it is crucial to bear in mind the ideological origins and functions of this kind of research. For two centuries, the liberal countries in the West have held to the ideology that merit is rewarded. Nowadays people would like to believe that those with ability and the will to work will succeed irrespective of sex, race or class origins. Scientists, who are members of a privileged elite that is predominantly male, white and middle or upper class, want to believe that they have personally earned their social privileges. Within this context, behavioral scientists see it as one of their professional tasks to investigate the intrapersonal differences between members of privileged and underprivileged groups so as to "explain" the differences in their positions and power within the society. These are truly "interesting" questions and questions for which it is possible to attract funding from the Ford and Rockefeller and Mellon Foundations. In deciding whether or not to work on these questions, scientists find that grant money, the feeling of doing important and useful work, and public attention lie one way, whereas early termination of a promising career or the obscurity of one's laboratory may lie the other way.

By downplaying the realities of the social situation that in fact severely limits access for members of underprivileged groups, scientists as a group deliberately or unwittingly support, and indeed help to generate, the ideology that allows the system from which they profit to run smoothly. Their work not only tends to fit into the dominant ideological perspective; it helps to create and solidify it by making it seem more "real."

In this introduction we have tried to offer a framework within which to think about the methodological pitfalls that beset research on the biological basis of sex differences. The papers that follow analyze in depth a number of the salient constraints and errors of the main lines of present-day research in this area.

Shorter versions of five of the papers were presented at a symposium at the annual meeting of the American Association for the Advancement of Science (AAAS) held in Washington in February, 1978. We thank the Association for Women in Science (AWIS) and Section X of the AAAS for sponsoring our symposium; Drs. Ethel Tobach and Betty Rosoff for including the manuscripts as part of their series of books, Genes and Gender; Florence Brauner and Ruth Manoff of the American Museum of Natural History for their meticulous editorial work; and the authors for their participation in the symposium and for helping us see this book through to completion.

Ruth Hubbard
Marian Lowe
August, 1978

NOTES AND REFERENCES

[1] M. L. Holbrook, *Parturition Without Pain: A Code of Directions for Escaping from the Primal Curse* (New York, Wood and Holbrook), 1875, In C. Smith-Rosenberg and C. Rosenberg, "The Female Animal: Medical and Biological Views of Woman and Her Role in Nineteenth Century America," *Journal of American History 60*, 335, 1973.

[2] S. S. Mosedale, "Science Corrupted: Victorian Biologists Consider 'The Woman Question,'" *Journal of the History of Biology 11*, 1-56, 1978.

[3] P. Geddes and J. A. Thomson, *The Evolution of Sex* (London, Walter Scott), 1889, p.267.

[4] J. McGrigor Allan, *Anthrop. Rev. 7*, ccxiii, 1869, quoted in Elizabeth Fee, "Nineteenth Century Craniology: The Study of the Female Skull," unpublished manuscript, 1977.

[5] S. J. Gould, "Morton's Ranking of Races by Cranial Capacity," *Science 200*, 503-509, 1978. In another paper Gould discusses the assumptions that guided Broca when studying sex differences using craniology, and he quotes Broca as follows: "We might ask if the small size of the female brain depends exclusively upon the small size of her body. Tiedemann has proposed this explanation. But we must not forget that women are, on the average, a little less intelligent than men, a difference which we should not exaggerate but which is, nonetheless, real. We are therefore permitted to suppose that the relatively small size of the female brain depends in part upon her physical inferiority and in part upon her intellectual inferiority." (S. J. Gould, "Women's Brains" *Natural History 87*,44-50, 1978.).

[6] J. S. Haller, Jr. and R. M. Haller, *The Physician and Sexuality in Victorian America* (Urbana, University of Illinois Press), 1974, p.50. S. Shields, "Functionalism, Darwinism and the Psychology of Women," *American Psychologist 30*, 739-754, 1975.

[7] D. L. Hall, "Biology, Sex Hormones and Sexism in the 1920's," in C. Gould and M. Wartofsky, *Women and Philosophy* (New York, G. P. Putnam), 1976, pp.81-96.

[8] A. D. Wood, "The Fashionable Diseases: Women's Complaints and Their Treatment in Nineteenth Century America," pp.1-23; R. Morantz, "The Lady and Her Physician," pp.38-53; in M. Hartman and L. Banner, *Clio's Consciousness Raised* (New York, Harper and Row), 1974.

[9] G. Dullea, "Vast Changes in Society Traced to Rise of Working Women," *New York Times*, November 29, 1977, p.1.

[10] A. W. Watts, *Nature, Man and Woman* (New York: Pantheon Books, 1958; Vintage Books, 1970).

[11] Thomas Kuhn, *The Structure of Scientific Revolutions*, 2nd edition, enlarged (Chicago: University of Chicago Press), 1970.

[12] C. Wright Mills, *The Marxists* (Hammondsworth, Middlesex: Penguin Books), 1963, p.12.

[13] H. Fairweather, "Sex Differences in Cognition," *Cognition 4*, 231-280, 1976.

[14] L. E. Tyler, *Psychology of Human Differences* (New York, Appleton Century Crofts), 1956, p.276.

[15] L. Eisenberg, "The *Human* Nature of Human Nature," *Science* 176, 123-128, 1972.

[16] I.K. Broverman, et al, "Sex-role Stereotypes and Clinical Judgments of Mental Health," *Journal of Consulting and Clinical Psychology 34*, p.1-7, 1970.

"UNIVERSALS" AND MALE DOMINANCE AMONG PRIMATES: A CRITICAL EXAMINATION

Lila Leibowitz

Northeastern University
Boston, Massachusetts

The genetic basis of behavioral traits is usually argued from the claim that a behavioral pattern is universal or nearly universal within a species. The notion that among humans sex roles are standardized and certain traits are universally those of one sex or the other is not a new one. Nor is the notion that such traits are genetically programed or "instinctive" a new one. In fact, for the century or so that anthropologists have been investigating other cultures systematically, these two intertwined notions have been tested time and time again. Much evidence has accumulated about some of the behaviors that were once considered part of the genetic heritage of one sex or the other, and, as a result, the hypotheses generally have been rejected because universals have not been found. Nevertheless, there are still a few anthropologists who see in the tremendously varied sex role assignments of men and women a pervasive pattern of male dominance, and some among them who regard what they perceive as male dominance to be biologically determined. The majority of social anthropologists, however, regard sex roles and statuses among humans as varied, learned, and the product of socioeconomic and cultural forces.

It is obvious to most students of other cultures that how labor is divided differs from one society or place to another. Who sews, or cooks, or hews wood, or draws water, or engages in market bargaining, or works in the fields, or produces the greater portion of subsistence foods are matters so varied as to defy simple sexual classifications. Societies also differ as to whether a biological mother is expected to nurse her infant or assume the major burden of caring for it, a fact which seems to surprise many Westerners who themselves belong to a tradition in which wet nurses and nannies are not long a thing of the past. Societies differ as to whether all husbands are men (West African peoples do not regard maleness as a prerequisite for husbandness). Societies differ as to whether spouses of either sex or both are taken one at a time or several at once; whether they live together or not; and whether they work with and for each other or not. More important in a discussion of sex roles and "dominance," and the degree — if any — to which they are biologically determined, are the different ways resources are controlled in different societies.

Data collected over the last hundred years show that there are quite a few variations as to who is in charge of collecting and distributing ordinary and special foods even in simple foraging societies. When women of particular kin or class or caste groups in more complex societies are in the position to allocate work or land or other valued goods to other members of their society, we are directly confronted with the problem of analyzing what sorts of control women and men exercise over the things which give people the power to negotiate decisions. Are Iroquois women who withhold the special dried foods men need for a war party exercising control over the domestic or political arena?[1] Is their veto power over men's decisions a form of dominance? Cultural variables in the control of strategic resources indicate that power relationships among humans, inter- and intra-sexual, cannot be reduced to the simple notion of "dominance" nor to its presumed biological components. As we shall see, oversimplification and vagueness as to what is meant by "dominance" are significant factors in the revival of the argument that male dominance is universal among humans, and among primate species generally.

Interestingly, the revival of the argument that sex roles among humans are genetically programed was stimulated by several scientists who do not specialize in studying human behavior but who did not regard this as a drawback for the purposes of producing popular books on animal behavior and human evolution and behavior.[2] Nonscientists and social scientists then hastened to get in on their act and benefit from the rich market they uncovered among Americans.[3] After questions were raised about the scientific validity, political bias and sexist prejudice of these popularized books, the argument was moved into the arena of "serious" scientific scholarship with the publication in 1975 of E. O. Wilson's *Sociobiology*, a text that received the prepublication treatment usually reserved for more readable books designed to reach general audiences.[4]

In a major work intended to lay the scholarly foundation for a "new" and innovative *science* of social behavior that he predicts will soon replace the softness of sociology and anthropology with the hardness of more rigorous biological subdisciplines, E. O. Wilson, a noted authority on social insects, stated that "aggressive dominance systems with males generally dominant over females"[5] are characteristic of the order Primates, the taxonomic order which includes monkeys, apes and humans. Coming as it does from a highly prestigious biologist in a text that addresses the issue of the evolution of social behavior, this statement has implications which are unavoidable for the conscientious student of human societies and cultures. If true, it implies that the observed range of human behavioral variability is either a departure from, or a conquest of, pre- or protohuman behavioral patterns, programs or predispositions. An important question, then, is, "Is this statement true?"

There are two ways in which I shall address the question of the validity of the proposal that dominance systems and male dominance over females characterize primate social behavior. First, I will examine whether the statement represents an accurate generalization drawn from the evidence which its author cites to support it. Secondly, I will examine the concept of "dominance" on which this author's analysis is built and which allows him to underwrite "scientifically" the notion that male dominance among humans is universal, biologically determined and hence difficult to overcome at best.

The text in which the above statement appears provides extended descriptions of the social arrangements of a number of primate species. The social arrangements that are described not only fail to justify the claim of widespread or near universal male dominance but are also somewhat confusing. Let me summarize some of the problems a reader encounters with regard to these descriptions.[6]

The mouse lemur is characterized as "an essentially solitary animal," although, we discover, mouse lemur females nest in groups. Evidently it is the *males* that are solitary. Whom or what they dominate in their solitary state is not clear. It appears that female nest groups are made up of mothers and daughters and their young who "displace" sons and brothers. "Dominant" males are characterized as those who manage to breed. Dominant males sometimes join females in their nests when the females are in estrus. Several males may join a nest when the females have passed out of estrous. The author notes, "the males evidently become more tolerant toward one another." Yet, who is becoming tolerant of whom is perhaps debatable, since all males usually are displaced to the outskirts of favored habitats.

Orangutans, the next species described, are designated as maintaining "nuclear groups," which consist of females and their young — occasionally accompanied by a usually solitary male. (In fact, the term matri-centered group seems more appropriate than "nuclear group" in this context, since the term nuclear family is used to describe the male/female/young family form among humans. But that's not the main issue just yet.) The author notes that "aggression within the society is quite rare, and nothing resembling a dominance system has been established in studies to date." Wilson cites a single instance of a female driving another female from a tree as the only clear episode of open hostility reported by observers. However, he states that males "probably do repel one another" because "a few pieces of indirect evidence suggest that such intra-sexual conflict does exist." That indirect evidence consists of the fact that male orangutans are much larger than females and have vocal pouches that make their calls extremely loud. By reading from morphology to behavior, the author presumes that large noisy males win out in "intrasexual conflicts" over females, though open conflicts between males were not observed. Reading from morphology to behavior is a dangerous business it turns out. Early on in the description of orangutans we find the statement, "As the orangs' unusual body form testifies, they are exclusively arboreal." Recent observers have learned, "The orangutan, studied in a rain forest in Indonesian Borneo, is not a tree dweller, contrary to popular belief but does almost 100 percent of its long-distance traveling on the ground."[7] Indirect evidence of body form probably tells us very little about the nature of intrasexual conflict between males, and certainly tells us even less about whether the normally solitary males are "generally dominant over females" with whom they rarely associate.

The dusky titis of the Amazon-Orinoco region and the white-handed gibbons of Malaya and Sumatra are next discussed in sequence. Dusky titis live in small groups which consist of a female, her young and a male. These mated pairs and their young are referred to as "one of the simplest familial forms of society." The titis, it is noted, share this societal form with, among others, the white-handed gibbons whose social arrangement is described as "identical to family." In the gibbon pair "the female plays an equal role in territorial defense and in precoital sexual behavior," though it is especially, but not always, the female who emits territorial calls. While "the mother takes care of the infant. . ." a lone gibbon male who allowed a small juvenile to adopt him and thereafter "carried the smaller animal in the maternal position during much of the day," indicates that "the male is also prepared to assume the role of the mother when she falls ill or dies."

Though it is tempting to regard the parenting capacities and pair mating arrangements of titis and gibbons in terms of the particular nuclear family form Americans have recently come to idealize, the extension of the term to nonhuman mating and nurturing arrangements violates the common practice of careful ethologists. It is important to note that in Wilson's descriptions the term "family" is used with reference to a form of animal grouping which resembles only one of the many kinds of groupings that are called family in human societies.[8] The use of the term "family" when referring to pair bond arrangements among nonhuman primates implies a biological basis for a, to us, familiar human social convention.

The mantled howlers who are described next are "of special sociobiological interest because a high level of individual tolerance permits the formation of large multimale societies." In addition, it is noted, they exhibit "the *unusual* circumstance of a species that appears to alternate between multimale and unimale organization and even has solitary males." The variability of howler social arrangements is clearly acknowledged. Conflict within troops is uncommon and almost never entails fighting. Not too surprisingly, we learn that in this species dominance orders are "weakly defined." Despite extensive observations by the seven researchers cited, "It has not yet been established whether the troops are age-graded-male, with one dominant individual controlling younger animals, or whether the troops contain multiple high-ranking males." The possibility that there might be no hierarchy is not entertained. No mention is made of behaviors that indicate that males are "dominant over females." The author simply assumes that in species where males are larger than females, as is the case among mantled howlers, the males must be dominant over the females. (The dangers of reading from morphology to behavior have already been pointed out.)

Ring-tailed lemurs also live in troops in which "fighting is rare." Yet their society is regarded as "aggressively organized." More notably we find that "adult females are dominant over males," which is "a reversal of an otherwise nearly universal primate pattern." While a linear hierarchy is observed among ring-tailed males, Wilson considers it "odd" that dominance in this hierarchy "seems to have no influence on access to estrous females." (Note that in the mouse lemur which is "solitary" and lacks a male hierarchy, "dominance" is attributed to males who have access to estrous females.)

It is hard to see on what grounds the claim for male dominance as a nearly universal primate pattern is being made, since up to this point in the argument Wilson has cited social organizations in which females seem to exclude males, social organizations in which males may or may not fight with each other, and social organizations which may or may not have male hierarchies, while no social organizations in which males determine or control female behavior have been described.

The other three primate species discussed in some detail are the hamadryas baboons, Eastern mountain gorillas, chimpanzees. All three are treated as giving evidence of "male dominance," although consistent criteria of dominance are not established. The hamadryas males of the small "single-male" units found in Ethiopia, herd and nip at females, effectively determining what the females will do. Among the peaceful gorillas who live in multimale groups, "most dominance interactions consist of a mere acknowledgment of precedence," which is to say that an animal, male or female, who gives up space to another is regarded by scientists as subdominant. In loosely-structured groups of chimpanzees we are told that "dominance behavior is well developed." Yet dominance behavior usually involves interactions which are "subtle" (again usually just giving way). "Overt threats and retreats are uncommon." Among chimpanzees we once again find that *"curiously. . .*(my emphasis) the dominance system appears to have no influence on access to females," who appear to solicit whom they please when they please. An estrous female who stopped grooming a dominant male to copulate with a subadult male exemplifies the situation.

Wilson's evidence to support the view that male dominance is universal among *all primates* is furthered with his description of the following human situation.

> Within a small tribe of Kung bushmen can be found in-dividuals who are acknowledged as the 'best people;' the leaders and outstanding specialists among the hunters and the healers. Even with an emphasis on sharing goods, some are exceptionally able entrepreneurs and unostentatiously acquire a certain amount of wealth. Kung men no less than men in advanced industrial societies generally establish themselves by their mid-thirties or else accept a lesser status for life. There are some who never try to make it, live in rundown huts and show little pride in themselves and their work. [9]

To set the record straight, the Kung are not tribal.[10] They live in camps of transient populations,[11] accumulate as little as possible to allow movement from camp to camp, and exchange and circulate materials and tools as well as food.[12] Camps include huts that are built — by the women, incidentally — at different times.[13] Leader-ship is ephemeral and task oriented, depending on who is in the camp and what has to be done.[14] There are no specialists other than shamans, some of whom are women.[15] The Kung have only recently become articulated with an entrepreneurial market economy, have no native category of "best people," and until recently, that is to say the past decade, discouraged competitiveness and pride.[16]

As we look over the evidence that Wilson offers, it becomes pretty clear that the generalization that primate males are usually domi-nant over females is arrived at by considering a minority of the species he describes as evolutionarily important and the majority of them as unimportant, to the point of ignoring some of them entirely. No statistical survey of the admittedly incomplete data on non-human primates is presented, although a series of tables summarizing some of the data on nonhuman primate social organization is offered. How reliable those data are with respect to "dominance" is another question, for clearly, there is considerable ambiguity and inconsistency just in the way the term itself is used.

Before examining the concept of dominance closely, however, let me point out that the order in which Wilson describes the social arrangements of the primates reflects his evolutionary model of "grades of sociality." Wilson explicitly rejects evolutionary models of behavior which stress either the biological relationships between primate species or focus on primate social organizational patterns as responses to ecological circumstances. Wilson's model is built on the notion that social evolution among primates involves a development from no-male, to one-male to multimale groups, but more significantly he assumes that social relationships evolve around males and male behaviors. Increases in male-to-male tolerance are at the heart of group development. That males are usually dominant over females in the no-male, one-male and multimale situations alike is expressly stated, although we've seen that this is clearly not the case. Furthermore, Wilson's notions about one-male-multimale groups are fuzzy; for example, one-male hamadryas groups are not truly comparable to the one-male groups of patas monkeys where the male is peripheral both socially and in space.[17] Males are emphasized as central in evolution because it is commonly assumed that whereas all females usually have infants, not all males have the same chance to breed. The idea that "dominant" males father more offspring than subdominant ones is so pervasive that it is said to be "odd" or "curious" when evidence to the contrary is found. Yet there is no doubt that "dominant" males do *not* have special sexual prerogatives in many, and perhaps most, species, not just in the ones Wilson regards as "curious" and "odd" (e.g., gorillas,[18] Japanese macaques,[19] cynocephalous baboons in forested areas,[20] chimps and others). Newer field studies clearly show that for many primates and in many situations social dominance is no guarantee of "success with the ladies." That the text ignores such data and rejects the examination of how primate social patterns are related to ecological settings and/or vary within species reflects the author's underlying premise that social behavior and social arrangements are genetically determined. Wilson's model of "grades of sociality" thus disregards evidence which suggests that there is an evolutionary trajectory involving increased reliance on learned and socially-transmitted behaviors in the primate order and makes more of the ill-defined notion of "dominance" than the data warrant.

To return to the uses and meaning of the term "dominance," the most thorough discussion of the concept of "dominance" and of primate behavior and social organization I have seen is that of Thelma Rowell.[21] Her review of the literature shows that whereas hierarchy and dominance-subordinance relationships have been considered the most important aspects of social behavior in animal groups, these concepts have been casually handled. Rarely have objective descriptions of social interactions been attached to statements about dominance. Furthermore, predigested, generalized observations make it impossible to compare studies by different observers who do not state clearly how they define and interpret the phenomenon they call dominance. Despite this obvious difficulty, there is widespread agreement that hierarchical relationships occur frequently among caged animals and are less clearly discernible or absent in noncaged groups. Studies that attempt to unravel the complex of factors usually associated with dominance have, therefore, been made on caged animals. The reasons for hierarchy in caged groups are complex. Such things as where food is placed, whether the animals were originally strange to each other, the age and prior experiences of the animals, and the nature of first encounters all play a part in the formation and maintenance of hierarchies.

In a 1970 paper, I. S. Bernstein[22] identified three dominance-related behaviors: aggression, mounting, and being groomed. For the study six species of monkeys were observed in groups living in large enclosures. Five of the six groups showed stable hierarchical relationships over several months with respect to the patterning of aggressive or agonistic encounters. In the sixth, a group of guenons, animals reversed their relationships several times during a year. Mounting relationships, and the hierarchies based on them, proved less stable than aggression hierarchies in all six species, and grooming relationships proved to be nondirectional and reciprocal. Bernstein found no correlations between the hierarchies obtained from the three kinds of relationships and concluded that they were not determined by a single social mechanism, were independent of one another, and were not necessarily determined the same way in each of the groups observed. In a later study, Bernstein and his associates[23] attempted to correlate aggressiveness and testosterone levels in an all-male group of rhesus monkeys and discovered that very high-ranking males—those who easily displaced all others —were neither very aggressive nor high in testosterone. A general correlation between aggressiveness and testosterone level, however, was found in lower-ranking males, who were under constant stress, leading Bernstein et al. to suggest that output of the hormone is determined by the animal's behavioral context, since the lowest-ranking, socially active males in this study had higher testosterone levels than males living in isolation. A key issue in the dominance-aggression-hierarchy equation appears to be stress.

Rowell examines the possibility that what is often discussed as "dominance" behavior is in reality "subordinance" behavior. Ultimately, she notes, the outcome of approach-retreat interactions are decided by the behavior of the potential retreater. Animals under stress tend to avoid interactions which may have unpredictable or negative results and hesitate to initiate them. Secure animals are far less cautious. Researchers tend to attribute high rank to those who approach others, whether or not they do so in an "aggressive" or agonistic manner, especially if they displace an animal that is avoiding them. Cages induce both high levels of stress and high interaction rates, which may be why dominance is so evident in the caged setting and why in these circumstances hierarchies become stabilized. In any event, "agonistic" hierarchies do not coincide with grooming or mounting hierarchies even in cages.

Rowell finds that there are several reasons for asking whether the concept of dominance is a useful one in discussing the evolution of primate social behavior: 1) "dominance hierarchies" are not consistent when they are determined for different types of behavior, so that the "top" animal is not the same in all situations; 2) among primates group behavior is rarely determined by coercion, so that a "top" animal does not in fact lead or control the group; 3) dominance has not been correlated with food-finding abilities or with danger avoidance; therefore, one cannot assume that dominance leads to significant survival advantages; 4) dominance by some may be an expression of the subordinance of others, which results from stress. This suggests that dominance is either an ephemeral or a long-lived result of situational settings rather than an independent trait, much less a genetically determined one; 5) the males who mate, or who do so most frequently, are by no means always dominant. Furthermore, although Rowell does not directly address the issue of whether male dominance is universal or nearly universal among primates, it is relevant that among monkeys she finds significant variations regarding the sex of those who constitute the core of particular social groups.

Jane Lancaster[24] also recorded some interesting observations that raised yet other questions about what is meant by male dominance. Her work on vervet monkeys shows that coalitions of females were easily formed against the top three males of the group she studied. If these offended some females by trying to monopolize a food source or by frightening an infant, even females of the lowest rank would band together to chase them. While a male's rank never changed as a result of such an encounter, his ability to bully others was curbed, and he learned to be very careful about frightening an infant. Several times Lancaster saw all the adult males leave an area when a nearby infant screamed. Though they in fact were not what had frightened the infant, their behavior clearly revealed their anxiety that the females would form a coalition against them.

It is hardly necessary to point out that Wilson's use of the term dominance as applied to primate societies reflects few of the caveats and cautions Rowell and Lancaster express about the behaviors sub-sumed under the notion of dominance. Furthermore, it is self-evident that he uses the term inconsistently; a male who breeds more than other males is defined as dominant on the basis of his breeding activities in a "solitary" species, yet in a more social arrangement a male who stands at the apex of a displacement hierarchy is called dominant, although he does not breed more than other males. Such inconsistencies — as well as the inaccuracies — in the use of a term do not inspire much confidence in Wilson's claim that one of sociobiology's virtues is that it will introduce behavioral and social scientists to the analytic rigorousness of biology.

It is eminently clear that the recent contention that male dominance is universal or nearly universal among primates is unfounded. This merely is a new version of one of many pseudo-biological arguments that are used to justify social arrangements in our society. The claim that these arrangements are found in our animal relatives suggests that these arrangements are the result of our genetic heritage. Now, as always, some researchers are trying to explain social traits of humans by attributing them to our innate biology. For the most part, however, such traits are no longer regarded as universal. Instead of talking about "instinctive, universal traits," researchers have been forced to talk about predilections or potentials for frequent or nearly universal traits. They therefore use vague notions such as "programmed potential" and "perceptual predisposition" to justify the conclusion that such traits are genetically determined. But like its other deterministic antcedents, this device ignores the history of changing human societies; just as it ignores the variety and variability of monkeys, it also ignores the fact that the *alternatives* to such frequent or nearly universal traits are also part of our human potential, and that "predispositions," "propensities" and "potentials" are developed in the contexts that favor them. The device of looking upon "near" universals as though they were genetic seems to be a new way of minimizing the need for analyzing the social contexts in which traits have developed. This is particularly insidious when these traits reflect social privileges for some people, and when they result from the unequal distribution of social privileges among the different social classes, races or the sexes. Incorrectly attributing to primates in general the male dominance that is nurtured and relished in our own society is not science; it is political propaganda.

NOTES AND REFERENCES

[1] M. Kay Martin and Barbara Voorhies, *The Female of the Species* (New York: Columbia U. Press), 1975, pp.225-227.

[2] See, for instance, Konrad Lorenz, *On Aggression* (New York: Harcourt, Brace and World), 1966, or Desmond Morris, *The Naked Ape* (New York: McGraw-Hill), 1968.

[3] Nonsocial scientist, Robert Ardrey, published *The Territorial Imperative* (New York: Atheneum) in 1966, His *African Genesis* (London: Collins) had appeared in 1961. Lionel Tiger's *Men in Groups* (New York: Vintage Books), 1970, was one of the first of the popular books by a social scientist to exploit this market.

[4] E. O. Wilson, *Sociobiology: The New Synthesis* (Cambridge: The Belknap Press of Harvard University Press). 1975.

[5] *Ibid.*, p.551.

[6] In the next few paragraphs of this paper a number of words, phrases and sentences are cited verbatim from the above text. Rather than providing the reader with a long list of "Ibid.s" and page references, I am noting here that these citations come from pages 514-546, a chapter entitled "The Nonhuman Primates."

[7] *Science News*, Vol.113, No.12, 1978, p.178.

[8] Lila Leibowitz, *Females, Males, Families: A Biosocial Approach* (North Scituate, MA: Duxbury Press), 1978, pp.6-9.

[9] Wilson, 1975, p.549.

[10] R. B. Lee, "The Kung Bushmen of Botswana," in *Hunters and Gatherers Today,* ed., M. Bicchieri (New York: Holt, Rinehart and Winston), 1972.

[11] R. B. Lee, "What Hunters Do for a Living, or, How To Make Out on Scarce Resources," in *Man the Hunter*, eds. R. B. Lee and I. DeVore. (Chicago: Aldine), 1968.

[12] L. Marshall, "Sharing, Talking and Growing; Relief of Social Tensions among Kung Bushmen of the Kalahari," *Africa*, Vol.31, 1961.

[13] F. Plog, C. J. Jolly and D. G. Bates, *Anthropology: Decisions, Adaptations and Evolution.* (New York: Alfred A. Knopf), 1976, p.486.

[14] *Ibid.*, p.425.

[15] R. B. Lee, Personal Communication.

[16] R. B. Lee, "Eating Christmas in the Kalahari," *Natural History*, Vol.77, No.10 (Dec.), 1969, pp.14-19.

[17] Thelma Rowell, *The Social Behavior of Monkeys* (Harmondsworth, England: Penguin Books, Inc.), 1972, p.63.

[18] George B. Schaller, *The Mountain Gorilla: Ecology and Behavior* (Chicago: University of Chicago Press), 1963.

[19] G. Gray Eaton, "The Social Order of Japanese Macaques," *Scientific American*, October, 1976, Vol.235, No.4, pp.96-107.

[20] Rowell, *op. cit.*, pp.46-66.

[21] *Ibid.*, pp.159-164.

[22] I. S. Bernstein, "Primate Status Hierarchies" in L. A. Rosenblum (ed.) *Primate Behavior*, Academic Press, 1970. Cited in Rowell, pp.161-162.

[23] R. M. Rose, J. W. Holaday and I. S. Bernstein, "Plasma Testosterone, Dominance Rank, and Aggressive Behavior in Male Rhesus Monkeys." *Nature*, Vol.231, pp.366-71. Cited in Rowell.

[24] Jane B. Lancaster, *Primate Behavior and the Emergence of Human Culture* (New York: Holt, Rinehart and Winston), 1975.

SOCIAL AND POLITICAL BIAS IN SCIENCE: AN EXAMINATION OF ANIMAL STUDIES AND THEIR GENERALIZATIONS TO HUMAN BEHAVIORS AND EVOLUTION

Ruth Bleier

University of Wisconsin
Madison, Wisconsin

Science is a cultural institution. While the structure of science has its edges pure and probing into the knowable unknown, its massive core, like all institutions, embodies, protects, and perpetuates the thoughts and values of those who are dominant in the society that produces it. To ignore this is to ignore the obvious. Scientists are human beings born into and molded by an insistent and obtrusively value-laden culture. The fact that scientists may be influenced in their work by what they *want* or *hope* or *believe* is generally acknowledged, as is evident by the self-imposition of experimental controls. We scientists consider such controls in biological and medical research to be essential to rigorous scientific investigation. It has even been possible to acknowledge that certain kinds of social biases affect scientific objectivity. The science historian Provine wrote in *Science* (1973):

> I am not condemning geneticists because social and political factors have influenced their scientific conclusions about race crossing and race differences. It is necessary and natural that changing social attitudes will influence areas of biology where little is known and the conclusions are possibly socially explosive...[1]

And socially explosive, indeed, are the implications of theories which deal with the origins and bases of sex differentiated roles in our society. They are made even more so by the potentialities of the woman's movement, of feminism, and of feminist critiques of our society's institutions, including its science. For unlike the issue of racism which is not likely soon to shake to its roots the structure of this overwhelmingly white, culturally middle-class society, the issue of sexism is already affecting every aspect in the spectrum of relationships between the two halves of our society. It speaks to questions of exploitation and dependence, love and hate, humiliation and fulfillment in every home, every office, every factory, every laboratory, every classroom.

Because of the explosive nature of the issue of sexism, because science is by and large a male institution, because sexist biases may be subtly incorporated into the very language and the implicit unspoken assumptions of scientific investigation, the task of providing a thorough and effective critique is formidable. But in addition to these inherent difficulties, feminists face the challenge of seemingly sober scientific truths coming from the pens of scientists who have earned stature and influence as a result of their pioneering and rigorous investigations in their own fields of study of insect, bird, or primate behavior. Undaunted and, indeed, protected by their well-earned scientific credibility, some of these wise men leap subtly to conclusions about the genetic-evolutionary inevitability of our contemporary human social relationships, bolstering the limited findings from their own field with carefully selected data gathered by colleagues in other fields, and, of course, ignoring all critical and contradictory findings. And there is no lack of interested parties to pick up on these kinds of stories and to publicize them.

For the past hundred years social and behavioral sciences have theorized about the nature and "proper" role of women, often sanctifying ancient beliefs and institutions which justified and maintained the economic, political, and social oppression of women. In recent decades, the biological sciences have attempted to approach questions of human "nature," behavior and social relationships, as well as their origins, by looking for the fundamental and basic, i.e., biological, mechanisms underlying human behaviors. This is done by dissecting or analyzing the responses of caged laboratory animals to various experimental manipulations or by observing the behaviors of animals in their natural habitats. Such biological research enjoys an authority derived from its presumed rigor and objectivity and from its efforts to analyze animal behavior which is seen by some to be uncontaminated by the effects of culture (as though animals do not have their own culture which varies with environmental changes). Extrapolations from this body of work to interpretations of human behavior constitute theories of biological determinism for human behavior and social organization. These have a curious tendency to explain and therefore justify most social evils as inevitable consequences of evolutionary strategies: such evils as the oppression and subordinate status of women, racism, classism, and war.

I should like in this paper to look at some specific examples of the forms in which sexist biases may appear in biological research and to indicate why certain assumptions are not valid. In general, biases may affect the language that is used, the assumptions (usually unstated) made, the questions asked, the controls used or not used, the data selectively used or ignored, the interpretations and conclusions drawn. With a skillful use of language and unspoken assumptions, any hypothesis can be confirmed, especially when investigators fail to state and consider possible alternative hypotheses. Further biasing occurs when the results of investigations of animal behavior are generalized inappropriately and uncritically to human behaviors. Underlying all the above, and common to most fields of investigation that involve humans and other animals, is a pervasive Western androcentrism; that is, a tendency to see in all of the animal world and in human evolutionary history the recapitulation of the behavior of the modern Euroamerican dominant male and of his social arrangements.

Sex Hormones and Behavior

My first examples of biases come from an area of laboratory research which has experienced extraordinary growth over the past two decades — the effects of hormones on behavior; more specifically, research on the effects of the sex hormones, estrogens and androgens,[2] on the fetal and newborn rodent's brain and on subsequent adult behavior. The part of the brain primarily involved in these studies is the hypothalamus, to the base of which is attached the pituitary gland. The pituitary is the body's principal endocrine organ, that by its hormones regulates the hormonal output from the other endocrine glands in the body — ovaries, testes, thyroid, adrenals, etc. The nerve cells (neurons) of the hypothalamus function not only like neurons in the rest of the brain but also like endocrine cells, since they too produce hormones, known as *releasing factors*, which regulate the hormonal output from the pituitary gland. Hypothalamic neurons are responsive not only to signals from neurons elsewhere in the brain but also to the levels in the blood of hormones from the endocrine organs — pituitary hormones, estrogens, progestins, androgens, and thyroid hormone.

The key findings in early work in this field were that: the hypothalamus is responsible, through its regulation of the pituitary, for cyclic production of hormones from the ovaries of the female rat and, therefore, for estrous cycles and cyclic ovulation; and androgens present at a critical period in the female or male fetal or newborn rat irreversibly suppress the capacity of hypothalamic neurons for cyclic hormonal activity.

These observations provided the intellectual framework for further work relating hormones, particularly androgens, to behavior, through their effects upon the developing brain. Thus, the initial findings were generalized to an important model for further experiments on the effects of androgens on mating and fighting behavior in rodents and primates, and it was concluded that the presence or absence of an androgenic "organizing" effect on the fetal or newborn brain determines the predominant pattern of these behaviors in the adult animal. These conclusions subsequently provided the basis for theories about the sexual differentiation of human brains, and of our social roles and behaviors.[3]

I shall later show that the extrapolations from animals to humans are unwarranted. In addition, even the basic model itself is flawed. *Estrogens* injected into female rats can produce the same so-called masculinizing effects that androgens do; i.e., suppression of estrous cyclicity, increased typically male mounting behavior, and increased fighting.[4] The reason appears to be that brain cells and many other cells of the body convert androgens to estrogens and, under some conditions, it appears to be the estrogens that evoke these effects.[5] Also, since a third sex hormone, progesterone (which, incidentally, is termed a "female" hormone) is a precursor to both estrogens and androgens, there is a very close chemical relationship among estrogens, androgens and progestins. All of them are present in women and men, though at different levels in different individuals and indeed in the same individual in different physiological and psychological circumstances. Furthermore, the effects in animals of sex hormone injections will vary depending upon the dosage used, the particular chemical form of the hormone, the species of animal, and the timing of the injection. Because of the metabolic conversions of one hormone into another, it is not really known, following an injection of any hormone, which is the final hormonal form producing the physiological or behavioral effect being measured. Finally, the basic rodent model of the so-called organizing effect of androgen on the developing brain cannot be extended even to nonhuman primates, for it has been demonstrated that androgens do not suppress cyclic functioning of the hypothalamo-pituitary system in male rhesus monkeys;[6] nor do they suppress it in female human and nonhuman primates who were exposed as fetuses to high levels of androgens.

The fact is that we have just begun to learn about these hormones, their receptors, and their mechanisms of action in the brain and in other tissues in all species, including humans, and we still understand very little about their effects. The effects on the developing brain (and entire organism) of the high maternal levels of estrogen and progestins to which *all* fetuses are exposed are only beginning to be investigated.[7] The complexity of these research problems suggests that this is not the time for scientists and others to use such data, which at best are preliminary and, hence, often contradictory as a basis for theories about the origins of human sex roles and behaviors. Even if the studies were conclusive, however, their relevance to human behaviors and social relationships would be highly questionable.

In studies of rodent mating behavior, it has been found that female rats which are given a large dose of androgens at birth, as adults more frequently mount other females or males, with attempts at intromission, than do untreated females.[8] (It is not unusual for untreated females of many species to mount other animals.) Conversely, male rats which are castrated at birth (deprived of androgens) and given estrogens as adults exhibit lordosis (the typical female mating posture with rump elevation) in the presence of other male rats.[9] These were the expected results and they are the ones that are most frequently quoted. But the picture is not so simple because it has also been found that in female rats, prepubertally administered androgens may *enhance* rather than abolish lordosis[10] whereas estrogens may *abolish* rather than enhance lordosis[11] and *increase* mounting activity with estrous females.[12] Furthermore, estrogen can inhibit rather than increase lordosis if given to a neonatally castrated male.[13]

The existence of this body of contradictory data has not prevented extrapolation to humans from the observations that estrogen-treated male rats *may* exhibit the typical female lordosis and androgen-treated female rats *may* exhibit an increase in typical male mounting. This has taken the ridiculous form of theories claiming a hormonal basis for human homosexuality[14] (i.e. that homosexuality is "caused" by exposure of the fetal hypothalamus to inadequate amounts of the sex-appropriate sex hormones or to an excessive amount of the so-called inappropriate or opposite sex hormones.) While some theories may be harmless, this one has led to irresponsible and radical efforts to "prevent" homosexuality by treating the human fetus through the mother with sex hormones or to "cure" it by brain surgery (lesioning the hypothalamus).[15]

From the point of view of scientific methodology, aside from the failure to consider contradictory data or alternative hypotheses, the attempt to extend the rat mating model to humans suffers from biased assumptions and simplistic thinking. It posits the existence in women and men of a set of stereotypical sexual behaviors that are reflexly unleashed by certain hormonal states, and implies that people (especially homosexuals) assume stereotyped sexual postures, equivalent to mounting or lordosis, that can be used as a measure of human sexuality. However sexual positions are only minor aspects of human sexual expression and choices of partners are based on complex factors that are culture dependent and multi-dimensional. Another biased assumption that has guided the search for hormonal abnormalities in homosexuals is that heterosexuality is the *only* biologically normal form of sexuality. A more objective assumption would be that bisexuality is the norm and that a person's sexual preferences are determined by the entire complex of experiences the person has had within the context of our patriarchal society, which officially condones and supports only monogamous heterosexual relationships. In fact, the vigor and emotionality with which heterosexuality is enforced makes one wonder about its biological normality.

The fact that only heterosexual mating results in reproduction is not relevant for the argument of the biological determination of sexual preferences, since most heterosexual intercourse is not for the purpose of reproduction. Nor does any one believe that homosexuality will become universal and threaten the survival of our species. There is no evidence whatsoever that homosexuality represents either a biological or an emotional aberration. In fact, the evidence supports a theory of multipotentiality of sexual expression. In many, perhaps the majority, of mammalian and also other species, males and females spend most of their time in same-sex groups, mostly ignoring the other sex. As stated before, among many animals, females mounting each other is not an unusual form of social activity. Heterosexual mating occurs only when the female is in estrus (heat) and solicits or permits it. Individual choice and hormonal state together determine the occurrence and the particular partners for mating. Among many, if not most, contemporary human cultures as well, women spend most of their time together (as do the men) and develop strong bonds based on kinship, support, cooperation and caring. With people, sexuality is not tied to hormones, i.e., to the menstrual cycle. Probably as powerful as the *hormonal* cues in the mating activity of most animals are the *cultural* cues for humans — in our society these are the insistent reinforcement of obligatory heterosexuality.

Male Aggression and Dominance

Studies of "aggressivity" in animals provide demonstrations of the use of biased and misleading language and of selective data to prove unspoken biased assumptions — in this case, that aggressivity, linked to high androgen levels, is the biological basis for universal and eternal male dominance and female subordinance. In most studies on "aggressivity," the behavior being observed and counted as one *measure* of aggressivity is the number of fighting encounters. The results and conclusions are discussed in terms of *aggressivity*, not of *fighting*, yet these terms, as commonly used in human contexts, are not at all synonymous. For example, following castration as neonates, male rats may fight less than do normal males, and following androgen injections as neonates, female rats may fight more often than do untreated females. The conclusion that is drawn is that *aggressivity* (however defined and wherever found) is androgen-dependent, more accurately, that in caged laboratory rats, *fighting behavior* may be influenced by androgens. A particular and specific measure of one sort of aggressivity becomes synonymous with aggression in general. But the term *aggressivity* is not value-free, objective or uniquely defined and, when used with reference to people, it is *not* synonymous with *fighting behavior*. Especially in primate research, the word is often used synonymously with dominance and is also socially endowed with a range of attributes from combativeness through assertiveness, independence, intelligence, creativity and imagination — mainly desirable human characteristics when associated with people, i.e., men who are leaders. So, by means of semantic flim-flam, such animal experiments are claimed to demonstrate that men are naturally (hence inevitably) dominant or superior to women because of inborn hormonal differences. This reasoning has been used by sociologist Steven Goldberg in his book, *The Inevitability of Patriarchy*, in which he rhapsodizes that without male aggressiveness we would not have "science, bureaucratic organization, industrialization and democracy."[16] Unfortunately, however, as Freda Salzman details in her paper in this collection, the theory of biologically determined aggressivity in males has appeal also for sober and serious investigators of the human personality, and even in their hands the definition of aggressivity undergoes some slippery transformations.

Since there are obvious limitations to comparing the behaviors of rodents and people, anthropologists have looked to present-day nonhuman primates for clues to the origins of human characteristics.

> The biological determinists assume that living nonhuman primates represent ancestral stages in human development and that primate behavior is a simpler version of human behavior, uncomplicated by cultural conditioning; that is, that human beings today act regularly in ways which may be traced backward to primate ancestors. But these primates are usually described as particular kinds of creatures — the males involved primarily with fighting to protect the territory which provides food resources, aggressive males dominating other males and all females. Such behavior and traits, according to this view, were inherited and further

developed by the hunting-gathering societies during the earliest, longest, and most decisive stage of human evolution, and in turn transmitted to us.[17]

The prototypical primate social unit is seen to consist of a dominant male, several females and young, with the male in the centrally important and enviable position of having first choice of sexual partners among his "harem" of females. That the term *harem* is used to designate a single-male troop of females is itself a revealing use of biased language and androcentric fantasy. It is a term that symbolizes the entire structure of social, economic, and political power relationships between the sexes; it evokes the image of Eastern potentates with their collection of dependent and powerless women. The use of this term to describe a single-male primate troop has, in effect, served as a *substitute* for scientific observations and investigations. Such is the force of language and of the subtle hypocrisies of science in the service of power. Jane Lancaster, however, gives us another and more rational view of polygynous troop structure and mating:

> For a female, males are a resource in her environment which she may use to further the survival of herself and her offspring. If the environmental conditions are such that the male role can be minimal, a one-male group is likely. Only one male is necessary for a group of females if his only role is to impregnate them.[18]

That some scientists and popularizers of science will carefully select data to confirm their pet theories is reflected in the fact that a favorite model for human behavior has been the troop of savannah baboons, in which the large aggressive male (the so-called alpha) is seen as defending the troop and its territory, dominating the hierarchy of other males and of all females, deciding troop movements, having first choice in food, sex, and grooming. Yet, when the same species of baboons lives in the forest, the *females* form the core of the troop, and determine when and where the troop moves; dominance and aggressive interactions are rare or nonexistent. When danger threatens, the first ones up the trees are the males, who are unencumbered by babies hanging on their fur. Furthermore, Rowell[19] has found that only when she captured and enclosed a troop of forest baboons did dominance hierarchies, aggressiveness, and competitiveness appear; that is, they were *learned* responses to specific situations.

The Problem of Animal Models

It is a common characteristic in the work of primatologists, ethologists, and sociobiologists — often magnified further in popularizations — that they choose a primate model for human behavior that reflects the behavior they are trying to "explain." The behavior they choose to explain reflects their beliefs or fantasies about relationships in human society and they then impose the language and concepts commonly used to describe human behavior not only upon their interpretations but even upon their observations of primate behaviors. The conclusion then becomes inevitable, for the entire structure is a self-fulfilling prophecy. The fact is that some primate model can be found to demonstrate *any* set of human characteristics or social interactions. As Lila Leibowitz describes in her discussion of universals and dominance in this volume and elsewhere [20] the data being accumulated by primatologists and anthropologists make clear that there exists no single pattern of aggressivity, dominance, troop defense, sex roles, sexual dimorphism, territoriality, competition or any other social behavior, either across or even within nonhuman primate species and human cultures[20].

Even if some sex-associated behaviors *were* found to be universal among all nonhuman primates or indeed among all mammalian species, generalizations to human behavior and social relationships would have to ignore 5 million years of exuberant evolutionary development of the human brain, which has resulted in a cerebral cortex quantitatively and qualitatively different from that of other primates. It is a cortex that provides for conceptualization, abstraction, symbolization, verbal communication, planning, learning, memory and association of experiences and ideas, a cortex that permits an infinitely rich behavioral plasticity and frees us, if we choose, from stereotyped behavior patterns. Furthermore, the human cortex constructs and transmits the body of ideas and values that constitute our culture and so liberates our human history from many pre-human modes of behavior. This determines the unique humanness of our behavior. Fallacious inductive reasoning leads from the stereotypic behaviors that constitute the survival repertory of certain species, such as ants or birds in their ecological niches, to the conscious, directed, intentional behaviors of humans — some of which may indeed be of dubious survival value. Humans have a vast capacity to learn and to create, as any examination of the variety of cultures throughout the world will reveal. Thus, not only is there no universal behavioral trait or repertoire among our closest relatives, the nonhuman primates, to study as a "primitive" prototype or precursor model for human "nature," there is no *human nature,* no universal human behavioral trait or repertoire that can be defined, *except* for our tremendous capacity for learning and for behavioral flexibility.

I find it remarkable that scholars, who in their personal lives, take for granted the uniqueness of each individual human consciousness and its sensitive molding by experience, exhibit no intellectual embarrassment when they invoke tiresome mystical concepts such as instincts or their more fashionable sociobiological equivalents — "pre-dispositions" or "tendencies" — to explain human behaviors. Yet, this is how they come to proclaim female nurturance on the one hand, or male dominance as manifested in patriarchy, territoriality or wars, on the other as innate biological drives that are based in evolution and hence, inevitable. If anything is innate and inevitable, it is our behavioral plasticity — the potential to develop in any of an infinite variety of ways in response to environment and culture.

I believe that one appeal of such theories lies in the fact that scientists and other intellectuals reared and trained in a liberal humanist tradition find the existence of oppression and war to be a heavy moral burden, especially if they feel obliged to assume responsibility for the human condition or to explain it. It is consistent with both the scientific and liberal traditions to be satisfied with analyses and conclusions that preclude a drive for change, let alone revolution: it is comforting to believe that things are as they are because they have to be so. Thus there is a predisposition to accept theories of biological determinism.

The Effects of "Male" Culture

There is another state of consciousness that I see affecting the validity of large bodies of research: the investigator's awareness of self (and those like him — and usually it is "he") as *universal,* as equivalent to humanity, viewing all others — the other sex, other classes, races, cultures and civilizations, other species and epochs — in the light and language of his own experiences, values and beliefs. He and his fraternity become the norm against which all the *others* are measured and interpreted.

Sheila Rowbotham[21] and Ruby Leavitt[22] have demonstrated that within this context of men's culture, everything relating to women comes out in footnotes to the main text, in sentences beginning *incidentally* or *by the way.* Take, for example, the remarkable statement by the anthropologist Marvin Harris in discussing the invention of agriculture, the so-called Neolithic Revolution: "Incidentally, there is little doubt that women rather than men made this momentous discovery."[23]

Again, in that context, the Man the Hunter theory of human evolution has been with us since the 1950's. We find prominent male anthropologists attributing all *human* cultural and intellectual evolution to courageous, strong, adventuresome, innovative *men,* banding with their fellows to hunt and bring home the meat to the dependent wife and babes. Tool- and weapon-making, language, and cooperation are said to have evolved from the hunting activities of early male hominids. "The biology, psychology, and customs that separate us from the apes — all these we owe to the hunters of time past."[24] In short, man evolved while woman incubated. Or let us hear E. O. Wilson, eminent sociobiologist whose area of expertise is *insect* behavior saying:

> In hunter-gatherer societies, men hunt and women stay at home. This strong bias persists in most agricultural and industrial societies and on that ground alone, appears to have a genetic origin. . .

And again,

> The building block of nearly all human societies is the nuclear family. . . During the day the women and children remain in the residential area while the men forage for game or its symbolic equivalent. . .[25]

And Frank Beach, eminent neuroendocrinologist, claims:

> Male genotypes that were above average in promoting
> those characteristics specifically related to effective
> performance of the hunter role were especially adaptive
> from the point of view of group survival.

(Beach states in another paragraph that these characteristics are
"certain emotional tendencies such as less fearfullness and greater
willingness to venture from the safety of the home base.") He continues,

> Within the female population, natural selection favored
> perpetuation and dissemination of those gene patterns
> which contributed most to behavior consonant with
> nonhunting, with gathering, with remaining near the
> home base.[26]

So, on the basis of a single proven sexual dimorphism, the bearing of
children by women, male anthropological and other schools of
thought have built an elaborate evolutionary mythology, which,
miracle of miracles, millions of years later still finds the little
woman in her proper and inexorable role at the kitchen stove.

But nowhere in such analyses do we find any of the data that
would suggest alternative reconstructions of human evolution. The
study, for example, of modern gathering-hunting peoples, such as the
!Kung (the closest models we have for the reconstruction of
Australopithecine and early hominid cultures), demonstrates that
plants, nuts, roots, and small animals gathered by women account
for 50 to 90 percent of the protein and caloric intake of the group,
and women control the distribution of the food they collect.[27] Their
gathering requires traveling up to 10 miles from the camp, and
women are as likely as men to be absent from home for days on end.
The child tenders are those who are not hunting or gathering on any
particular day. This basically egalitarian situation among the !Kung
has begun to change only in the past 10 years as they have settled
near the more Westernized Bantu, and, as has happened throughout
the world, wherever contact has occurred with modern Western
technologists, men were the ones who were introduced to more

modern agricultural methods and who therefore have come to control the means of production.

It is of more than trivial sociological interest that the collection of papers from the symposium on hunter-gatherer societies that demonstrated the central importance of women gatherers to the subsistence and social organization of these peoples is entitled *Man the Hunter*.[28] And it is fairly breathtaking to see that in the volume by the same editors devoted to studies of the !Kung and other Kalahari hunter-gatherers, Washburn can say in the foreword: "...it may well be that it was the complex of weapons-hunting-bipedalism which accounts for the evolutionary origin of man."[27] As the creator of this theory, he is not likely to be moved by contradicting facts.

But as Zihlman and Tanner point out, the archeological evidence for the butchering of large trapped animals with flake stone tools is only 0.5 million years old and for the hunting of large animals with weapons (an elephant with a spear between its ribs) is only about 100,000 years old, a tiny fraction of the 5 or so million years of human evolution.[29] Fossil evidence suggests that in early times, most hunting involved the gathering of small animals and required ingenuity and speed, not strength, or it involved the butchering of large animals that were sick, dead, or mired.[29] Furthermore, Australopithecene and other early hominid fossil findings indicate that these ancestors had the large, worn, grinding molars and premolars of plant-eating animals and did not have the long tearing canines of carnivores. These data,together with observations of the eating and predatory habits of modern chimpanzees and baboons, suggest that early hominids depended upon plant foods and that meat was an occasional fortuitous addition. There are other fossil observations that are relevant to a discussion of the postulated singular role of hunting in the evolution of human characteristics. The human brain developed from the 500 cubic centimeter capacity of the brain of early hominids of 4 to 5 million years ago to the 1200 cubic centimeters of *Homo erectus* of 100,000 years ago. Taking 1450 cubic centimeters to be the average size of the modern human brain, this suggests that 83 percent of human brain development occurred during the 98 percent of human evolutionary time that passed before the hunting of large animals developed into a full-scale social effort about 100,000 years ago.

Most anthropologists and paleontologists agree that the tremendous leap forward in hominid brain growth and development occurred in response to, and in concert with, the skeletal anatomical changes that made possible upright bipedal locomotion, which, in turn, freed the hands and broadened the visual range. Further evolution of the hands for a fine precision grip and increased eye-hand and body co-ordination required a larger brain which, in turn, made possible both increased motor skills and the development of language and communication skills. As Ruth Hubbard has written[30] "It is likely that the evolution of speech has been one of the most powerful forces directing our biological, cultural and social evolution... [and] that the more elaborate use of tools and the social arrangements that go with hunting and gathering came in part as a consequence of the expanded human repertory of capacities and needs that derive from our ability to communicate through language. Speech probably predates the evolution of hunting by hundreds of thousands of years."

Thus, there is no evidence that hunting could have been a force driving the evolution of early upright hominids 3 to 5 million years ago, since it appears to be part of our much more recent history. Since it is likely that early hominid woman, like her present-day gatherer-hunter sisters, was the gatherer of plant and small animal food as well as the carrier of nursing babies, it is probably she who invented the earliest tools (diggers and levers), baby slings, food containers and choppers as well as agriculture, agricultural tools and pottery.[31]

If we are to look at the growing bodies of data from primatology and anthropology with eyes and intellects unclouded by conventional wisdoms, biases, and fantasies, we may be able to begin to learn about the origins of human social organization. But only then can we begin to see Woman the Gatherer as an important source and central participant in human cultural evolution.

To summarize and conclude, science, like any cultural institution, reflects the values, beliefs, and interests of those who make it. Two of the closely related biases that most seriously afflict and restrict fields of investigation of animal and human social behaviors and cultural evolutionary theory are parochial androcentrism and sexism. Thus, studies of animals are interpreted to confirm the preconceptions that motivated the studies: that humans (by which is meant *men*) are the *killer* apes, driven by their "instincts" (i.e., genetic-hormonal heritage) for aggressivity, lust, and territoriality to violence and war. Yet, the fact that the human species has not only survived but proliferated everywhere on the face of the earth would seem to prove that unbridled aggressivity and territoriality did not lead its early representatives where it logically would have — into the evolutionary dead end of intraspecies conflict, competition, and annihilation. Rather the best evidence, as well as logic, suggest that the qualities of sharing and cooperation that characterize present-day gatherer-hunter societies were always basic and necessary features of evolving human societies.[31] [32] It is probably only in very recent history, the last several millenia, in which we had first the growth of settled agricultural economies succeeded by economies that depended on industry and the accumulation of property, that these developed states and governments which rule large groups of people, and for which wholesale intraspecies murder in the form of war, has become an accepted feature of human life.

While the notion of the killer ape has been dramatized and popularized as best-selling science fiction, at least as insidious though less dramatic has been the distortion of science through theories of a biological basis for male "superiority," to explain the social, political, and economic dominance of men in most cultures.

Since survival of evolving hominid species probably required that every woman from puberty on be either pregnant or nursing (which contemporary gatherers do for the entire 4-year period between births), the consequent restrictions on her freedom of movement and agility and also the need to protect nursing infants from danger probably placed women at an economic disadvantage with the advent of large animal hunting about 100,000 years ago. That is, in ecological settings requiring heavy dependence on meat, women would no longer have the central role in food production that they had in predominantly gathering societies, and would lose control both of means of production and of the product which confers status on those who have and distribute it.[32] It is not difficult to imagine the subsequent codification, enforcement and magnification of this sexual inequality (in the distribution of wealth and power) by the development of a body of rituals, traditions, institutions, laws and beliefs — the course of development and the particular content of this body of culture being unique to each evolving society. With the change from a gathering-hunting nomadic way of life to settled farming-herding communities about 10,000 years ago, it became possible to accumulate possessions and property. Wives with their ability to provide male offspring as heirs to a man's possessions became an essential part of the property of men, just as their sisters and daughters became part of a man's wealth to be exchanged with other propertied men for material goods and for the purposes of kinship alliances.[33] The particular set of marriage and kinship arrangements, incest and other taboos, differ from society to society but all have or had in common the implicit ownership by men of women's bodies and lives. Thus, these cultural expressions and enforcements of female subordinance and male dominance assumed a life and evolution of their own and have survived all subsequent transformations of modes of production to present-day industrial societies, and, in fact, become even stronger and more sophisticated as their original biological justification becomes more remote in time and relevance.

What might have begun as a division of labor, based on a straightforward sexual/reproductive difference, now finds reinforcement and embellishment with "scientific" theories and assumptions about innate differences in intelligence, abstract thinking, logical reasoning, emotional stability, independence and passivity, visual spatial perception, stamina, etc. But sexual stereotypes, like all others, are stifling. They are now anachronistic and continue to be a basis for social organization only because they serve the interests of those who benefit from the *status quo*, not because they reflect some innate biological traits.

In general, my criticism of some areas of biological research lies within a recent tradition of feminist analysis which views the public spheres of human culture as entirely male throughout written history until the present beginning revolution in thought kindled by feminist critics and creators. The namers and the sayers have been men.[34] Women have been absent from language, history, philosophy and theology, anthropology, sociology, psychology — either omitted entirely from consideration (as irrelevant appendages) or constructed as creatures too dark, different, primal, mysterious, chaotic to include in rational and disciplined discourse. But not only have we been *missing* from the intellectual inquiries and productions of civilized men, we have been silent. This is not, some say,[35] because we have no voice but because that voice, our existence, our consciousness, our passion, our significance, has been repressed. But women together, with our different modes of inquiry and criticism, our different tongues and nationalities, have begun to break our collective silence. We are "demystifying and deconstructing" traditional (male, or, as French women critics name it and say it, phallologocentric) scholarship and forms of discourse. All that remains to do is to write and speak ourselves into being: to construct a new language, a new scholarship, a new knowledge that is whole.

NOTES AND REFERENCES

[1] W. B. Provine, "Geneticists and the Biology of Race Crossing," *Science*, 182, 1973; pp.790-796.

[2] By way of definition it should be noted that *androgens* constitute a family of steroid hormones, including testosterone, produced by testes, ovaries and adrenal glands, but in largest amounts by testes. *Estrogens* are a family of steroid hormones produced by the same three glands but in largest amounts by the ovaries. *Progestins* are produced mainly by the ovarian corpus luteum but are present also in males in small amounts as part of the metabolic pathway of androgens.

[3] For a more extended treatment of some issues and studies discussed in this paper, see R. Bleier, "Myths of the Biological Inferiority of Women: An Exploration of the Sociology of Biological Research," *Univ. Mich. Papers in Women's Studies*, Vol.2, 1976, pp.39-63.

[4] C. Doughty and P. G. McDonald, "Hormonal Control of Sexual Differentiation of the Hypothalamus in the Neonatal Female Rat," *Differentiation*, Vol.2, 1974, pp.275-285. P. G. McDonald and C. Doughty, "Androgen Sterilization in the Neonatal Female Rat and Its Inhibition by an Estrogen Antagonist," *Neuroendrocrinology*, Vol.12, 1973/1974, pp.182-188. R. E. Whalen, C. Battie, and W. G. Luttge, "Anti-Estrogen Inhibition of Androgen Induced Sexual Receptivity in Rats," *Behavioral Biology*, Vol.7, 1972, pp.311-320.

[5] K. H. Ryan, F. Naftolin, V. Reddy, F. Flores, and Z. Petro, "Estrogen Formation in the Brain," *American Journal of Obstetrics and Gynecology*, Vol.114, 1972, pp.454-460. J. Weisz and C. Gibbs, "Conversion of Testosterone and Androstenedione to Estrogens in Vitro by the Brain of Female Rats," *Endocrinology*, Vol.94, 1973, pp.616.

[6] F. J. Karsch, D. J. Dierschke, and E. Knobil, "Sexual Differentiation of Pituitary Function: Apparent Differences Between Primates and Rodents," *Science*, Vol.179, 1973, pp.484-486.

[7] B. H. Shapiro, A. S. Goldman, A. M. Bongiovanni, J. M. Marino, "Neonatal Progesterone and Feminine Sexual Development," *Nature*, Vol.264, December, 1976, pp.795-796.

[8] G. W. Harris and S. Levine, "Sexual Differentiation of the Brain and Its Experimental Control," *Journal of Physiology*, Vol.191, 1965, pp.379-400.

[9] S. Levine, "Sexual Differentiation: The Development of Maleness and Femaleness,"*California Medicine*, Vol.114, 1971, pp.12-17.

[10] F. A. Beach, "Male and Female Mating Behavior in Prepuberally Castrated Female Rats Treated with Androgen," *Endocrinology*, Vol.31, 1942, pp.673-678.

[11] S. Levine and R. Mullins, "Estrogen Administered Neonatally Affects Adult Sexual Behavior in Male and Female Rats," *Science*, Vol.144, 1964, pp.185-187.

[12] P. Sodersten, "Mounting Behavior in the Female Rat During the Estrous Cycle, After Ovariectomy, and After Estrogen or Testosterone Administration," *Hormones and Behavior,* Vol.3, 1972, pp.307-320.

[13] H. H. Feder and R. E. Whalen, "Feminine Behavior in Neonatally Castrated and Estrogen-Treated Male Rats," *Science,* Vol.147, 1965, pp.306-307.

[14] G. Dörner, W. Rohde, F. Stahl, L. Krell, and W. G. Masius, "A Neuroendocrine Predisposition for Homosexuality in Men," *Archives of Sexual Behavior,* Vol.4, 1975, pp.1-8.

[15] H. F. L. Meyer-Bahlburg, "Sex Hormones and Male Homosexuality in Comparative Perspective," *Archives of Sexual Behavior,* Vol.6, 1977, pp.297-325. As is true for all other fields of study, women have been omitted from psychology as legitimate sentient and human subjects of inquiry and investigation. Thus speculations and theories about, as well as research in, homosexuality have been primarily concerned with males. However, this is an instance, where it has been to our advantage to be overlooked in view of the quality of most work in this area.

[16] S. Goldberg, *The Inevitability of Patriarchy* (New York: Morrow), 1973.

[17] R. R. Leavitt, *Peaceable Primates and Gentle People Anthropological Approaches to Women's Studies* (New York: Harper and Row), 1975.

[18] J. B. Lancaster, *Primate Behavior and the Emergence of Human Culture* (New York: Holt, Rinehart and Winston), 1975.

[19] T. Rowell, *Social Behavior of Monkeys* (Baltimore: Penguin), 1972. T. Rowell, "The Concept of Social Dominance," *Behavior Biology,* Vol.11, June, 1974, pp.131-154.

[20] J. B. Lancaster, *Primate Behavior and the Emergence of Human Culture* (New York: Holt, Rinehart and Winston), 1975; E. Leacock, "Women in Egalitarian Societies," in: *Becoming Visible: Women in European History,* eds. Renate Bridenthal and Claudia Koonz (Boston: Houghton Mifflin), 1977; R. R. Leavitt, *Peaceable Primates and Gentle People: Anthropological Approaches to Women's Studies (New York: Harper and Row), 1975; L. Leibowitz, "Perspectives in the* Evolution of Sex Differences," in: *Toward an Anthropology of Women,* ed. Rayna R. Reiter (New York: Monthly Review Press), 1975; L. Leibowitz, *Females, Males,* **Families: A Biosocial** (North Scituate, MA: Duxbury Press), 1978.

[21] S. Rowbotham, *Woman's Consciousness, Man's World* (Baltimore), 1973.

[22] **R.R. Leavitt, Peacable Primates and Gentle People: Anthropological Approaches** *to Women's Studies* (New York: Harper and Row), 1975.

[23] M. Harris, *Culture, Man and Nature* (New York: Thomas Crowell), 1971.

[24] S. Washburn and C. Lancaster, "The Evolution of Hunting," in: *Man the Hunter,* ed. R. B. Lee and I. DeVore (Chicago: Aldine), 1968.

[25] E. O. Wilson, *Sociobiology: The New Synthesis* (Cambridge: Harvard University Press), 1975.

[26] F. Beach, "Human Sexuality and Reproduction," in: *Reproductive Behavior,* ed. W. Montagna and W. Sadler (New York: Plenum), 1974.

[27] R. B. Lee and I. DeVore, eds., *Kalahari Hunter-Gatherers* (Cambridge: Harvard University Press), 1976.

[28] R. B. Lee and I. DeVore, eds., *Man the Hunter* (Chicago: Aldine), 1968.

[29] A. Zihlman and N. Tanner, "Gathering and the Hominid Adaptation," in *Female Hierarchies* (New York: Harry Frank Guggenheim Foundation Third International Symposium), April, 1974.

[30] R. Hubbard, "Have Only Men Evolved?" in *Women Look at Biology Looking at Women,* eds. R. Hubbard, M.S. Henifin and B. Fried (Cambridge, MA: Schenkman Publishing Co.), 1979.

[31] S. Slocum, "Woman the Gatherer: Male Bias in Anthropology," in *Toward an Anthropology of Women,* ed. Rayna Reiter (New York: Monthly Review Press), 1975; N. Tanner and A. Zihlman, "Women in Evolution, Part I: Innovation and Selection in Human Origins," *Signs: Journal of Women in Culture and Society,* Vol.1, 1976, pp.585-608.

[32] R. E. Leakey and R. Lewin, *Origins* (New York: E. P. Dutton), 1977.

[33] G. Rubin, "Traffic in Women: Notes on the 'Political Economy' of Sex," in *Toward an Anthropology of Women,* ed. Rayna Reiter (New York: Monthly Review Press), 1975.

[34] A. Rich, *Of Woman Born* (New York: W. W. Norton), 1976.

[35] E. Marks, "Women and Literature in France," *Signs,* Vol.3, 1978, pp.832-842. C. G. Burke, "Report from Paris: Women's Writing and the Women's Movement," *Signs,* Vol.3, 1978, pp.843-855; The quotation in the last paragraph is from Marks's paper.

AGGRESSION AND GENDER:
A CRITIQUE OF THE NATURE-NURTURE
QUESTION FOR HUMANS

Freda Salzman

Department of Physics
University of Massachusetts
Boston, Massachusetts 02125

Are males more aggressive than females? If so, is the greater aggressiveness of males due to innate factors, or is it a result of boys being brought up differently from girls? This is, of course, an example of the nature-nurture question which has been resurfacing in various forms in the past decade.

The particular question of whether males are innately more aggressive than females has received considerable attention in recent years. An alleged biologically based sex difference in aggression plays an extremely important and central role in a succession of highly publicized theories, from the "naked apery" theory of human behavior,[1] to the most recent and highly acclaimed theory of sociobiology of E. O. Wilson.[2] According to these theories, aggression is the primary means by which human dominates human; hence, it is argued, in male-female relations, the innately more aggressive men dominate women. This thesis is then used to explain women's subordinate position in the family and the workplace. (This same model is also used to explain the highly stratified, hierarchical structure of our society, based mainly on class, race, and sex, and the correspondingly differentiated distribution of status, wealth, and power.)

Sex difference in aggression is just one of a number of complex social behavior traits that have been claimed to be genetically or biologically based. I will focus on this particular question for two reasons. First, the claimed biologically based sex difference in aggression plays a fundamental role in a number of theories, including that of sociobiology which, in particular, has gained a great deal of legitimacy and respectability. For this reason alone, we ought to know more about the validity of this claim. And second, the methodological difficulties encountered in this problem are characteristic, and thus illustrative, of those which arise in the general question of establishing a genetic or other biological basis for differences in complex social behavior traits (within the normal range).

In order to assess the validity of the claim for biologically based sex differences in aggression, we need to know rather explicitly what actual studies have been performed and exactly what has been demonstrated. As a representative example, I look at the evidence cited to support the claim made by Eleanor Maccoby and Carol Jacklin in *The Psychology of Sex Differences* that males are more aggressive than females and that "the male's greater aggression has a biological component."[3] Since this book is probably regarded as the most important and comprehensive current work on sex differences, I will examine in detail the authors' evidence *pertaining to humans*. I will demonstrate that their arguments have such serious methodological flaws as to invalidate their conclusion. This analysis reveals many of the complexities and difficulties of the general nature-nurture question and I use it as a case study to discuss the general problem of establishing genetic or biological determinants of complex social behavior traits within the normal range. (Obviously, infants born with major anatomical or physiological abnormalities may fall outside the established norms of behavior.) Exactly the same criticisms that I raise have been leveled against attempts to establish a genetic basis for class and race differences in achievement, the most publicized of which is the IQ controversy.[4]

Ambiguity of the Concept of "Aggression" and its Social Significance

The first requirement in any study that attempts to establish a "genetic or biological component" of a trait is that the trait be defined and quantified with utter precision so that it can be dealt with in a scientific manner. "Aggression" is one of the murkiest concepts in the behavioral sciences, and many behavioral scientists are guilty of using ambiguous and questionable definitions. (See Bleier, this volume.) The term "aggression" is defined in the dictionary to mean "unprovoked attack" or "physical assault." In the lay language, the term encompasses a whole range of attributes, ranging from anti-social behavior, such as in violent crime, to combative behavior, such as in warfare, to highly regarded social traits, such as competitiveness and dominance, which are admired and deemed to be necessary in leaders and other achievers in our society. Rather than carefully avoiding the ambiguity resulting from the use of this term in the lay language, some behavioral scientists seem to exploit it.

An example of the association of the term aggression with violence occurs in the myth of the "criminal chromosome," in which it was claimed that males with an additional Y chromosome committed violent crimes more often than did normal XY males. The explanation given was that the Y chromosome in normal males contributes to aggressive tendencies, so that XYY "supermales," since they have two Y chromosomes, are prone to "superviolence." It is now known that there is no correlation of the XYY trait with higher rates of violent crimes.[5] This is consistent with the fact that, as I will demonstrate here, there is no credible evidence that the usual XY males are innately more aggressive than are females.

An example of the use of the term aggression at the other end of the spectrum, as a desirable trait necessary for success, is given by Wilson in a popularized version of sociobiology, in which he states:

> As shown by research recently summarized in the book, "The Psychology of Sex Differences," by Eleanor Emmons Maccoby and Carol Nagy Jacklin, boys consistently show more mathematical and less verbal ability than girls on the average, and they are more aggressive from the first hours of social play at age 2 to manhood. Thus, even with identical education and equal access to all professions, men are likely to play a disproportionate role in political life, business and science.[6]

We find all this ambiguity—and more—in the use of the term "aggression" reflected in *The Psychology of Sex Differences*. At the outset of the section on "Aggression,"[7] the first section in the chapter titled "Power Relationships," the authors state that the word "aggression" refers to a loose cluster of actions and motives that are not necessarily related to each other, but the central theme of which is "the intent of one individual to hurt another." In discussing aggressive behavior in animals, they frequently refer to actual fighting behavior, that is, one animal physically attacking another. They close the section with the statement that aggression in the *male* may express itself in a number of ways other than in "interpersonal hostility" (which was not the original definition of "aggression") such as in "dominance hierarchies" and "competition." In a concluding section of the text entitled "Dominance, Leadership and Vocational Success,"[8] the authors suggest that there might be an aggressive element in the "iron-fisted tycoon" and the "aggressive" salesman. They indicate that male dominance and leadership in the past has been linked to aggression. Further, they state:

> We must leave it to the reader's judgment to estimate how often the "killer instinct" is involved in achieving success in the business or political world. Clearly, it sometimes is, and in these cases there will be a smaller number of women than men who will have the temperament for it . . . We believe we see a shift toward more *nonaggressive* leadership styles in high-level management, but at the moment this is speculation. [emphasis added][9]

Thus, we find Maccoby and Jacklin extending their use of the term "aggression" in a direction which is identified with success and leadership in our culture. Still other definitions of the term aggression used in the book are discussed below.

The Evidence:
I. Sex Hormones and Aggression

Let us now turn to the evidence pertaining to a biological basis for sex differences in aggression in humans cited by Maccoby and Jacklin. They state that "Aggression is related to levels of sex hormones, and can be changed by experimental administrations of these hormones."[10] The first study of humans that the authors cite is one by Anke Ehrhardt and Susan Baker, and it is used primarily to support a current model of the sexual differentiation of the human brain by prenatal sex hormones, which has received considerable attention and has been fairly widely publicized in academic circles, particularly through the book *Man and Woman, Boy and Girl* by John Money and Anke Ehrhardt.[11] The direct evidence for this model of sexual differentiation has been derived entirely from studies of animals, primarily rats. In this model there is a critical period of development, which for rats occurs just after birth, but for humans is believed to occur before birth, during which time the sex hormones organize or "program" male and female brains in ways that predispose females and males to respond differently to certain stimuli and hormones. In rats it is found that in the critical period, the testes of males secrete testosterone (one of the androgen hormones, frequently referred to as "male" hormones because they occur in higher levels in males), which is said to organize a "male" brain. This "organization" is irreversible, leads to an essentially acyclic (male) pattern of sex hormone production and determines characteristic stereotyped "male" behavior, which in rats means fighting and mounting (copulatory) behavior. The brain of a rat that is not exposed to high levels of testosterone (or any other closely related hormone, such as progesterone, one of the two sex hormones which occur in higher levels in females and which therefore are referred to as "female" hormones) during the critical period, regardless of the chromosomal sex, develops an irreversibly "female"-organized brain. This leads to a cyclic (female) pattern of ovarian function and ovulation, with an estrous period ("heat"), when stereotypical "female" behavior is displayed, which includes assuming the lordosis position (elevation of the rump) in the presence of mature males. Chromosomal females exposed during the critical period to high levels of androgen-like hormones (or even to estrogens — see Bleier, this volume), however, develop stereotypic male behavior.[12]

With this animal model in their minds, Ehrhardt and Baker studied the behavior of 17 chromosomal XX females who were exposed while still in the womb to unusually high levels of androgens due to the malfunctioning of their adrenal glands, a disorder which is referred to as the adrenogenital syndrome (AGS). These fetally "androgenized" AGS girls were born with male external genitalia in various stages of development and subsequently underwent corrective surgery to make them look like normal females. All the girls required continued treatment with cortisone to help correct the adrenal malfunctioning. The behavior of these girls compared with that of their eleven normal sisters was claimed to be "masculinized" in that "they much more often preferred to play with boys; they took little interest in weddings, dolls, or live babies, and preferred outdoor sports."[10] These girls were found to initiate more fights, but this difference was not found to be statistically significant.

Maccoby and Jacklin themselves indicate some of the problems with this study. The girls were on continued cortisone treatment, which although designed to bring their cortisone concentrations up to normal levels, might have led to unknown side effects. Furthermore, the behavioral information came from interviews with the mothers. The evidence is so weak that Maccoby and Jacklin conclude their discussion with the statement, "It is primarily in their consistency with animal work in early administered hormones that the findings with human subjects become especially compelling."[13] This statement is absurd — and "bad" science, given the substantial problems, with the animal studies themselves, the further problems of identifying and defining analogous behavior in animals and humans, and the considerable variation of sex-differentiated behavior from species to species that is seen even for closely related species (see Leibowitz and Bleier, this volume). Elizabeth Adkins, in a recent, comprehensive survey of studies of the effects of the early administration of hormones to animals, discusses in greater detail many of the difficulties with this work and with its use as a basis for identifying hormonally based human sex differences.[14]

In a later account of their work,[15] Ehrhardt and Baker noted that they reexamined the fighting behavior of the AGS girls. They also obtained more detailed data on this behavior, but it had not as yet been analyzed. As a result of their review, the researchers stated that it was not clear whether these fetally androgenized females differed in any way from their normal sisters (and mothers, who were also included in the "control" group), in respect to childhood fighting behavior. This result is in accord with a whole series of studies of fetally androgenized girls which consistently show that these girls are *not more aggressive* than other girls, particularly with respect to fighting behavior.[16] Their behavior has been described as being "tomboyish," but not to such an extent as to make them conspicuously different from other girls.

Critiques of these studies have pointed out that the whole effect can quite plausibly be explained by environmental factors resulting from parental ambivalence toward daughters born with malelike genitalia, regardless of what mothers report in retrospect their attitudes to be. Perceptions of behavior can differ markedly from observed behavior, and studies show that parents quite unconsciously treat boys and girls differently from early infancy, as discussed below. Ehrhardt reported that in their study, in six cases, marked external anatomical deformities (malelike genitalia) were corrected shortly after birth; in seven cases, clitoral enlargement was corrected in the second and third year; and in four cases, the surgical correction was performed later in life.[17] A number of other criticisms of these studies have been made, as well.[18] For example, although Ehrhardt and Baker call the normal sisters and mothers of the AGS girls a "control group," it is highly questionable whether this is the case, as noted by Adkins in her critique of studies of AGS girls.[14] The only kind of control group which would be meaningful would be one in which the members were in every way similar to the AGS girls except that the controls would not be exposed to high levels of prenatal androgens. This means, for example, that the control girls would also have had to be born with penises or clitoral enlargements which were surgically corrected, have had the same kind of stressful childhood and received the same extensive and prolonged medical attention, and have been raised uncertain of their physical femininity and role as potential mothers because of their biological problems. Obviously such real "controls" are virtually impossible to find.

The primary evidence offered to support the claim that prenatal hormones sexually differentiate the human brain with respect to socially significant behavior traits are studies of fetally androgenized females, such as the one just discussed. Extensive critical reviews demonstrate that these studies are highly questionable.[14] Therefore, to date, there is no substantive evidence that prenatal hormones sexually differentiate the human brain with respect to social behavior.[18 19] In fact, there now appears to be some evidence counter to this model of the human brain. It turns out that

- 77 -

the irreversible sexual differentiation of the hypothalamus, which in rats and some other species has been shown to govern the cyclic production of hormones in females — a cycling which is suppressed by early exposure to androgens in males — apparently does not apply to humans (or to non-human primates). In male rats, and in females exposed to a high level of androgens during a critical period after birth, the hypothalamus irreversibly loses the capacity for the cyclic hormone response typical of the female; male rats castrated at birth display a capacity for the cyclic hormone response. In monkeys, however, it has been demonstrated that the hypothalamus of a normal adult male retains the capacity for responding in the cyclic (female) pattern.[20] Furthermore, studies show that the cyclic female pattern of hormone production can occur in women (and in female non-human primates) who have been exposed to high levels of androgen before and after birth, as is the case with women with the adrenogenital syndrome.Thus, the model of the sexual differentiation of the brain, at least with respect to this specific physiological function (the capacity for the cyclic hormone response) *cannot* be extrapolated from rats to humans or to non-human primates.

As further evidence that aggression is related to levels of sex hormones, Maccoby and Jacklin claim that "Male hormones increase aggressive behavior when they are administered postnatally even without prenatal sensitization."[21] But the authors indicate that there is little research in humans relevant to this question. They then cite the study of boys with the adrenogenital syndrome, also included in Ehrhardt and Baker's work, who were exposed to greater than normal prenatal levels of androgens. Ehrhardt and Baker found that boys with the adrenogenital syndrome are *not* behaviorally different from their normal brothers. However, in spite of the fact that this study shows *no* effect, Maccoby and Jacklin conclude with the statement that "It is an open question whether, among human beings, variations in the amount of testosterone present prenatally are associated with individual differences in aggressive behavior during the growth cycle."[21]

Finally, Maccoby and Jacklin claim that "More aggressive males tend to have higher current levels of androgens."[21] The only study of humans they cite is one by Kreutz and Rose in which blood plasma testosterone levels were measured in a group of 21 young men in prison. Men with higher testosterone levels had supposedly committed more "violent and aggressive crimes" during adolescence. Included in their list of crimes, however, are not only such things as armed robbery, assault, murder, and attempted murder, but also "escape from prison." In fact, the *only* "violent and aggressive crime" committed by the prisoners with the top and third highest levels of testosterone was "escape from institution."[22] Moreover, the prisoner with the second highest testosterone level had committed the crimes of armed robbery and assault, but whether the assault occurred separate from, or during and as a consequence of, the armed robbery is not clear.

The best that can be said for the Kreutz and Rose study is that it is inconclusive. Further, Kreutz and Rose point out that the mean value and the range of levels of testosterone of the prisoners is close to what is normally found for males of this age group. They also note that there are many young men, without criminal histories and who have not committed violent acts, with levels of testosterone as high as those of the prisoners who had committed the supposedly more aggressive crimes in adolescence. Even the studies that claim that testosterone levels are correlated with aggression have not demonstrated a causal relationship.[23] In fact, several studies show that the experiential situation can affect the levels of testosterone. For example, testosterone levels are found to be suppressed during periods of psychological stress that have no obvious connection with aggressive behavior.[24]

The Evidence:
II. Observations of the Behavior of Children

Another type of evidence cited by Maccoby and Jacklin is that males are more aggressive than females in all human societies for which they claim evidence is available.[25] In the text, they discuss two cross-cultural studies, both involving observations of children. In one, the playground behavior of children was observed in three societies, the United States, Switzerland, and Ethiopia. Aggression was defined as "pushing or shoving without smiling," and boys were found to engage more frequently in this type of behavior than girls in all three societies. The other cross-cultural study cited is Whiting and Pope Edwards's analysis of observations of children's behavior in six cultures. In all the cultures studied it was found that direct physical assault of one child upon another was rare and no statistically significant statement could be made about sex differences. Boys were found to engage in more rough-and-tumble play, to exchange more verbal insults, and to counterattack more frequently if aggressed against.

Maccoby and Jacklin ignore a considerable body of evidence of societies in which neither sex is aggressive, such as the Tasaday and the pygmy societies, and, what is particularly noteworthy, reports of the lack of aggressiveness of children in socialist China, even in urban neighborhoods.[26]

The final evidence cited by Maccoby and Jacklin is that the sex differences in aggression are found early in life — "as early as social play begins — at age 2 or 2½" — before, they claim, there is evidence of differential socialization for aggressive behavior.[25][27] Since, in fact, 2-year-olds do not usually go around physically assaulting others with "intent to hurt," Maccoby and Jacklin introduce a new concept, "attenuated forms of aggression," which they define to mean "mock-fighting" and, "aggressive fantasies."[27][28] Mock-fighting, in turn, is defined to be "rough-and-tumble play." Therefore, what is actually observed in 2-year-olds is rough-and-tumble play, which then becomes synonymous with the term aggression, itself!

According to the authors, one of the best established sex differences is the much greater incidence of rough-and-tumble play among boys.[28] Although this is certainly a far less sweeping generalization than the claim that sex differences in aggression are found in all cultures, even this claim is highly questionable. In fact, one of the important findings of the Whiting and Pope Edwards analysis is that in societies in which boys and girls have similar tasks and responsibilities, there are fewer sex differences between boys and girls, including rough-and-tumble play.[29] For example, in a town in Kenya where young boys take care of younger siblings and help with domestic chores, the magnitude of the sex differences is one of the smallest in the cultures studied, and the decrease is due primarily to the decrease in "masculine" behavior in boys. Girls from the ages 3-11 years actually engage more frequently in rough-and-tumble play than boys, but the difference is not statistically significant. Similar findings are reported for another town in East Africa, and by Carol Ember for a town in Western Kenya.[30]

Moreover, Whiting and Pope Edwards also find that in a New England town where young girls are not involved in child care, sex differences again are small and 3-6-year-old girls engage more in rough-and-tumble play than do boys, but not significantly so.

Maccoby and Jacklin's claim that sex differences in aggression — meaning "rough-and-tumble play" — are found early in life and that there is no evidence of differential socialization up to that time is rather surprising in terms of their own discussion. Turning to the chapter entitled "Differential Socialization in Boys and Girls," we find in the section on "Total Parent-Child Interaction" that they explicitly note that studies of parental behavior find a "consistent trend for parents to elicit 'gross motor behavior' more from their sons than from their daughters," and that this takes several forms.[31] One study found that parents are more likely to handle male infants "roughly and pull their arms and legs vigorously." In another study, "fathers reported engaging in more rough-and-tumble play with their sons than their daughters." According to the authors, a central theme of these studies appears to be that "girls are treated as though they are more fragile than boys." In another section, fathers are reported to be extremely disturbed if there are indications of "femininity" in their sons and want their daughters "to fit their image of a sexually attractive person."[32]

These results are entirely consistent with other studies which indicate that boys and girls are treated differently from the time social interactions begin following birth.[33]

How then do the authors support their claim that there is no differential socialization with respect to aggression? The authors' remarkable claim rests upon a further metamorphosis of the meaning of "aggression" to mean such things as destructive acts, insolence or being "cheeky" toward the parents, anger or hostility to parents, and hitting back when attacked.[34] It is not surprising to learn that many parents in our society respond similarly to their sons and daughters with respect to this particular list of traits. Thus, again, we are faced with the problem arising from the ambiguity in the meaning and use of the term "aggression." In fact, Maccoby and Jacklin themselves indicate that they are aware of a problem because they note that there *are* reports of sex-differentiated attitudes of parents with respect to aggression, in that fathers tend to worry if a son is "unaggressive," meaning here "unwilling to defend himself."[35] They explain these apparently contradictory findings with the statement that "some of the inconsistency in the data no doubt stems from different understandings of what is implied by 'aggression.'"[35]

Interaction of Genes and Environment

Unfortunately, Maccoby and Jacklin do not mention until the next to the last page of the textual material the interesting results found by Whiting and Pope Edwards and by Ember that young boys involved in caring for younger siblings were less aggressive than boys who did not have such responsibilities.[36] Moreover, their conclusion is that "the process of caring for children *moderates aggressive tendencies*" [emphasis added]. Elsewhere, they state that social shaping of sex-typical behavior is related to "certain sex-linked *predispositions*" [emphasis added].[37] The concepts of "tendencies" and "predispositions" are used by some behavioral scientists to indicate that while a range in the expressions of a supposedly biologically based trait is possible, a particular expression is more probable. Hence, these terms are misleading because they imply that certain types of behavior will develop under "natural" conditions, and that to change or modify them one has to impose special "unnatural" conditions. In particular, these words imply that for a period of time, or on a continuing basis, conscious social pressure would have to be applied to maintain the "unnatural" state needed to curb the "natural" tendency of the organism. What Whiting and Pope Edwards and Ember point out, however, is just the reverse: where boys and girls are brought up similarly, with the same task assignments, particularly that of child care, their behavior also is quite similar, and this includes rough-and-tumble play. This situation can hardly be viewed as one in which boys are subjected to special social pressures which help to "tame" them, so "moderating their aggressive tendencies."

The concepts of "tendencies" and "predispositions" misrepresent the way in which genes and environment interact. From modern genetics we learn that genes as well as environment are involved in every aspect of the development of an organism, including behavior. All the observable aspects of an organism, called the phenotype, depend upon the genetic nature of the organism, called the genotype, *in interaction with* the environment in which the genotype develops. The present model is dynamically interactive: genes and environment continually affect the ways in which the organism develops, and do so non-linearly by affecting the organism's ways of responding to gene-induced as well as environmental changes. In humans there is the additional feature that we ourselves can shape or influence our environment in major, significant ways.

At the present time, geneticists study the interaction of genes and environment in plants and experimental animals, such as fruit flies, by experimentally varying a specified and well-controlled environment for a given genotype and seeing what phenotype results. In general, as the environment is varied, different phenotypes can result. The set of possible phenotypes for a given genotype is called the *norm of the reaction.* The expression of some traits for a given genotype can be extremely sensitive to environmental conditions and may vary considerably as the environment varies; the expression of other traits may remain fairly constant as the environment is varied. There is no way to select a particular environment as "natural" for a particular genotype.[38] Moreover, the norms of reaction for different genotypes can differ considerably from each other.

As an example of the interaction of genetic, biological and environmental factors, consider the trait "rough-and-tumble play" which is one of the behaviors frequently observed in extremely young children. I discussed earlier the evidence that environmental factors can be significant from infancy. Further, it is known that environmental factors can have developmental effects on infants and on young children. We are usually aware of the possibility that an infant who is treated roughly and played with more vigorously may develop differently psychologically than one who is treated very gently and regarded as though it were "fragile." But what is sometimes not considered is that such differential treatment can also produce physiological and physical differences, for example, with respect to neural and muscular development, and adrenal responses (see also Star's paper in this volume). Children who have been played with more vigorously, and who have developed physically and psychologically as a result of such play, may, in turn, engage more in such play with their cohorts, and may enjoy it more.

Hence, scientific studies which attempt to determine genetic or hormonal bases for differences between individuals or between groups would require rearing children from birth in precisely defined and controlled environments. In studying sex differences in behavior, particular caution would be required in the experimental design to eliminate not only sex-differentiated treatment by child-rearers, but also sex-biased interpretations by the observers.[39] In effect, the experimental situation would have to be designed to be "sex-blind" to both rearers and observers. At present, it is highly improbable that this can be achieved.

Furthermore, in the case of complex social behavioral traits, the controlled laboratory experiments cannot begin to replicate the actual multitude of interactions which even infants and very young children experience. Thus, even if we performed superbly controlled laboratory experiments, we would only learn from them about genetic and hormonal determinants of behavior in specific and extremely limited, artificial situations. At present, there is no theory that enables us to extrapolate from such "experimental" environments to our actual, complex society. From studies in a particular environment, we still would have *no* basis for deciding how resistant or plastic a trait of a given genotype would be to changes in the environment.

Conclusion

What conclusions can be drawn from the above discussion? First, from the many studies of young children there is a sizeable body of evidence that indicates that actual hitting and fighting behavior does not occur among children cross-culturally. Also, there are indications that in societies in which there is a higher incidence of rough-and-tumble play among boys than among girls, girls and boys are brought up differently. Furthermore, since "rough-and-tumble play" has not been shown to be synonymous with behavior with the "intent to hurt," none of the studies support the claim that boys are more "aggressive" than girls. Moreover, there are no studies to show that children, who at age 2 or 2½ years have the highest incidence of rough-and-tumble play, are those who as adults either commit violent crimes or are particularly successful in the academic, corporate, or political sphere. Therefore, it is thoroughly irresponsible — and dishonest — of behavioral scientists to denote as "aggression" (or such euphemisms as "attenuated forms of aggression," "mock-fighting," or, as it is also sometimes called, "mock-aggression") the cluster of behaviors observed in very young children that is called rough-and-tumble play, given the highly charged meaning and the strong connotations — both positive and negative — that the words "aggression" and "aggressive" have in our society.

Finally, the theoretical and practical difficulties of establishing the *biological* determinants of behavior in humans within the normal range, appear to be insurmountable at this time. Now and for the foreseeable future, one should be extremely skeptical of studies that purport to establish a biological basis for group differences in complex human social behaviors that lie within the normal range.

NOTES AND REFERENCES

1 R. Ardrey, *African Genesis* (Dell), 1961, *The Territorial Imperative* (Dell), 1966, and *The Social Contract* (Dell), 1970; K. Lorentz, *On Aggression* (Bantam), 1966; D. Morris, *The Naked Ape* (Dell), 1967 and *The Human Zoo* (Dell), 1969; L. Tiger, *Men in Groups* (Vintage), 1970; and L. Tiger and R. Fox, *The Imperial Animal* (Dell), 1971.

2 E. O. Wilson, *Sociobiology: The New Synthesis* (Harvard Univ. Press), 1975.

3 E. Maccoby and C. Jacklin, *The Psychology of Sex Differences* (Stanford Univ. Press), 1974, p.360.

4 N. Block and G. Dworkin, eds., *The IQ Controversy* (Pantheon), 1976; The Ann Arbor Science for the People Editorial Collective, *Biology as a Social Weapon* (Burgess), 1977, Section "Race and IQ," pp.19-55.

5 H. Witkin, et al., "Criminality in XYY and XXY Men," *Science*, 193, 13 August 1976, pp.547-555; R. Pyeritz, et al., "The XYY Male: The Making of a Myth," Ann Arbor Science for the People Collective, eds. (cited above, note 4), pp.86-100.

6 E. O. Wilson, "Human Decency is Animal," *New York Times Magazine*, 12 October 1975, p.50.

7 E. Maccoby and C. Jacklin, (cited above, note 3), pp.227-247.

8 *Ibid.*, pp.368-371.

9 *Ibid.*, p.368.

10 *Ibid.*, p.243.

11 J. Money and A. Ehrhardt, *Man and Woman, Boy and Girl* (Johns Hopkins Univ. Press), 1972.

12 For further details of this model of the human brain, see *Ibid.*

13 E. Maccoby and C. Jacklin, (cited above, note 3), p.244.

14 E. K. Adkins, "Genes, Hormones, and Gender" in G. Barlow and J. Silverberg, eds., *Sociobiology: Beyond Nature-Nurture* (American Association for the Advancement of Science, in press).

15 A. Ehrhardt and S. Baker, "Fetal Androgens, Human Central Nervous System Differentiation, and Behavioral Sex Differences," in R. Friedman, R. Richart and R. Van de Wiele, eds., *Sex Differences in Behavior* (Wiley), 1974, pp.33-51.

16 J. Money and A. Ehrhardt, (cited above, note 11), pp.10,99,103; J. Money and M. Schwartz, "Fetal Androgens in the Early Treated Adrenogenital Syndrome of 46 XX Hermaphroditism: Influence on Assertive and Aggressive Types of Behavior," *Aggressive Behavior*, Vol.2, No.1, 1976, pp.19-30.

17 R. Friedman, et al., (cited above, note 15), p.82.

18 R. Bleier, "Myths of the Biological Inferiority of Women," *The University of Michigan Papers in Women's Studies*, Vol.2, 1976, pp.39-63; K. Grady, "A Review of Man and Woman, Boy and Girl," *Science for the People*, Vol.9, No.5, 1977, pp.36-38; E. K. Adkins, (cited above, note 14); "Discussion: Effect of Hormones on the Development of Behavior," in R. Friedman, et al., (cited above, note 15), pp.77-84.

19 F. Salzman, "Are Sex Roles Biologically Determined?," *Science for the People*, Vol.9, No.4, 1977, pp.27-32, 43.

20 F. Karsch, D. Dierschke, and E. Knobil, "Sexual Differentiation of Pituitary Function: Apparent Difference Between Primates and Rodents," *Science*, 179, 1973, pp.484-486.

21 E. Maccoby and C. Jacklin, (cited above, note 3), p.246.

22 L. Kreutz and R. Rose, "Assessment of Aggressive Behavior and Plasma Testosterone in a Young Criminal Population," *Psychosomatic Medicine*, Vol. 34, No.4, 1972, pp.321-332.

23 A. Kling, "Testosterone and Aggressive Behavior in Man and Non-Human Primates," B. Eleftheriou and R. Spott, eds., *Hormonal Correlates of Behavior* (Plenum), 1975, pp.305-323.

24 L. Kreutz, R. Rose, and J. Jennings, "Suppression of Plasma Testosterone Levels and Psychological Stress," *Archives of General Psychiatry*, Vol.26, 1972, pp.479-482.

25 E. Maccoby and C. Jacklin, (cited above, note 3), p.242.

26 For a detailed critique, see: B. Chasin, "Sociobiology: A Sexist Synthesis," *Science for the People*, Vol.9, No.3, 1977, pp.27-31.

27 E. Maccoby and C. Jacklin, (cited above, note 3), p.352.

28 *Ibid.*, p.237.

29 B. Whiting and C. Pope Edwards, "A Cross-Cultural Analysis of Sex Differences in the Behavior of Children Aged Three Through 11," *The Journal of Social Psychology*, Vol.91, 1973, pp.171-188; also in A. Kaplan and J. Bean, eds., *Beyond Sex-Role Stereotypes* (Little, Brown and Co.), 1976, pp.188-205.

30 C. Ember, "Feminine Task Assignment and the Social Behavior of Boys," *Ethos*, Vol.1, No.4, 1973, pp.424-439.

31 E. Maccoby and C. Jacklin, (cited above, note 3), pp.307-311.

32 *Ibid.*, pp.328-329.

33 J. Rubin, F. Provenzano, and Z. Luria, "The Eye of the Beholder: Parents' Views on Sex of Newborns," *American Journal of Orthopsychiatry*, Vol.44, No.4, 1974, pp.512-519; also in A. Kaplan and J. Bean, eds., (cited above, note 29), pp.179-186; C. Seavey, P. Katz and S. Rosenberg Zalk, "Baby X, The Effect of Gender Labels on Adult Responses to Infants," *Sex Roles*, Vol.1, No.2, 1975, pp.103-109; S. Goldberg and M. Lewis, "Play Behavior in the Year-Old Infant: Early Sex Differences," *Child Development*, Vol.40, 1969, pp.21-31.

[34] E. Maccoby and C. Jacklin, (cited above, note 3), pp.323-327.

[35] *Ibid.*, p.326.

[36] *Ibid.*, p.372.

[37] *Ibid.*, p.275.

[38] For further discussion, see R. C. Lewontin "Biological Determinism as a Social Weapon," in The Ann Arbor Science for the People Editorial Collective, (cited above, note 4), pp.6-18; F. Salzman, (cited above, note 19). For a more detailed discussion, see R. C. Lewontin, "The Analysis of Variance and the Analysis of Causes," *The American Journal of Human Genetics,* Vol.26, No.3, 1974, pp.400-411, reprinted in N. J. Block and G. Dworkin, eds., (cited above, note 4), pp.179-193.

[39] J. Condry and S. Condry, "Sex Differences: A Study of the Eye of the Beholder,"*Child Development,* Vol.47, 1976, pp.812-819.

SOCIOBIOLOGY AND BIOSOCIOLOGY: CAN SCIENCE PROVE THE BIOLOGICAL BASIS OF SEX DIFFERENCES IN BEHAVIOR? *

Marian Lowe and Ruth Hubbard
Boston University   *Harvard University*
Boston, Massachusetts   *Cambridge, Massachusetts*

Biological Determinism and Social Change

Of the recent challenges to contemporary social structures, redefinitions of sex role behavior pose probably the greatest threat. They raise the possibility of change in all the ways in which the two halves of the population relate to each other and to social institutions. Indeed, some changes are already here; major economic and social effects are occurring as women, particularly those with small children, enter the work force in large numbers.[1]

Anxieties over the real and imagined consequences of such major social rearrangements have led to a number of efforts to restrict the equal access of women (and of others who have been hitherto excluded) to the work force. The battles over the Equal Rights Amendment and the Bakke decision are two well-known examples. These anxieties have provided the impetus for major reexaminations of the origins of sex role differences and of the significance of these differences for the theory and practice of social equality between women and men. Not surprisingly, most of this work has been done through academic research, since in our society academics are primarily responsible for providing a suitable basis, scientific if possible, for the ideological constructs that explain and rationalize the social order and make it politically palatable.

However, considering the profound economic and social importance of this issue, it has not been as widely debated as one might expect. Instead, a number of biologists and social scientists have

*This paper is based in part on an earlier paper by M. Lowe, "Sociobiology and Sex Differences," *Signs,* Vol.4, 1978, pp118-125.

adapted some of the various forms of biological determinism that reappeared as part of the anti-Civil Rights backlash of the late 1960s and that recently have provided the basis for Jensen's and Herrnstein's controversial (and incorrect) claims regarding the heritability of IQ.[2] The more recent efforts have attempted to demonstrate scientifically a biological basis for some of the behavioral sex differences that are readily observed in our society — nurturance, competitiveness, aggressivity and others. Most respectable among these has come to be sociobiology, the attempt to turn social behavior into biology by proving that it has an evolutionary, hence genetic, component.[3]

In order to do this, analogies are drawn between what are said to be universal aspects of human social behavior and the ways in which various female and male animals —insects, fishes, birds, mammals — relate to each other. The assumption is that similar behaviors imply a common biological basis, with a common evolutionary history, and this is taken to prove genetic causation of the behavior.

The other contributors to this volume discuss many methodological flaws that not only haunt *inter*specific comparisons of behaviors among animals and humans, but that make some of the *intra*specific studies suspect in animals as well as in people. We agree with their critiques. We would, in addition, like to emphasize that, although in every question we ask of nature, our social, economic, and political biases become unavoidable components of our answers, this is particularly true when we operate in areas where the very notion of a "controlled experiment" is wishful thinking. The study of the origins of human behavioral sex differences is, of course, just such an area.

In the last year it has become fashionable in some academic circles to say that sociobiology is based on overly simplistic evolutionary reductionism and naive attempts to generalize about behavioral similarities in different species. New voices, such as sociologist Alice Rossi, have begun to argue that the division of labor in which men's principal function is said to be productive and women's reproductive (a division that has held sway for only the last few centuries and even then only in parts of the globe) should not be seen as a crude dichotomy into female and male behavior completely determined by genes. Instead, a model is proposed which suggests that "on the average, because of their evolutionary history, men and women differ in their *predispositions* to care for infants, and in their *ability* to learn those caretaking skills" [our emphases]. These differences are said to appear as a consequence of recently discovered effects of sex hormones on the organization of the developing, embryonic brain in some animal species.[4] These effects, it is argued by analogy, may program the brain differently in women and men.[5] In typical, hair-splitting academic jargon, this is described as an interactive, "biosocial" treatment as distinct from "sociobiological" causation. In our view, the theories are virtually indistinguishable and both are flawed in that they try to substitute unique, biological causes for the many complex interactions that determine divisions of social and economic roles and power in Western, capitalist societies.

Why is it worth demonstrating the flaws in these theories? Why not let time and new work take care of them? E. O. Wilson argues that even if his interpretations regarding the innateness of behavioral sex differences were correct and if these differences did in fact impose differential limits on performance by women and men, this "could not be used to argue for anything less than sex-blind admission and free personal choice."[6] Yet a conviction that there are relevant, innate sex differences would surely have profound effects on the struggle to attain "sex-blind admissions and free personal choice." At the very least it would affect the goals and tactics. At worst, it could lead to abandoning the effort and so become a self-fulfilling prophecy. Rossi argues that inherent differences exist and should affect both personal choice and social policy. She believes that women should not confuse difference with inequality; that illusions of equality will produce masculinized women and alienated children; that a healthy society must exploit women's innately superior nurturing abilities.[4]

Such theories are not trivial intellectual exercises; they have profound implications for social policy. Hence it is imperative that we examine them for their scientific merit, and since they are being widely publicized, also for their possible political impact.

The Scientific Questions

The question of scientific merit can be summed up simply. None of these theories can distinguish, in a way that satisfies the minimum requirements of the scientific method, the extent to which a given behavior is caused by environment or by biology. To show why this is so we will concentrate on the basic assumptions of sociobiology and of biosociology.[7] Readers should note, however, that many of the methodological problems we discuss are not unique to these theories.

Sociobiology: Sociobiology is the study of the biological basis of behavior. It attempts to show that human social institutions and social behavior are the results of biological forces acting through prehuman and human evolution. The theory is based on Darwin's theory of evolution through natural selection, which sociobiologists claim to extend and amplify. The central assumption is that any behavior that has some genetic component is *ipso facto* adaptive; that is, that it came to be inherited (was "selected") precisely because it "worked." A major difficulty in *applying* the theory to real life situations involves the main point under discussion — how to determine whether a behavior has a genetic component. The sociobiologists themselves point out some of the difficulties involved. Wilson speaks of

> ...the considerable technical problem of distinguishing behavioral elements that emerge... independently of learning and those that are shaped at least to some extent by learning. Where both processes contribute, their relative importance under natural conditions is extraordinarily difficult to estimate.[8]

This considerably understates the difficulty. Even for animals, it is often hard to distinguish genetic and environmental effects, since even the lowliest animals are able to learn. For people, with our large capacity to learn and to systematize what we have learned in our culture, the difficulty may be insurmountable. Even where one cannot doubt that genetic factors exist, as in the sex differences in some physical characteristics, recent work has shown that environmental influences can be profound and, indeed, can overshadow biology. For example, in comparing arm strength of women and men

Several researchers have concluded that much of this

difference is the result of society's encouraging the average man to be more active than the average woman. They feel that the social influences are so great that inherent physiological differences in strength cannot yet be estimated.[9]

Sociobiologist David Barash has noted another difficulty:

> ... just because something is adaptive does not mean that it must be biological even if it accords with **sociobiologic prediction. But it is equally true that being** cultural does not require that it cannot be biological *as well* ... it seems scientifically appropriate ... to employ theory to make testable predictions. [However] If reality accords with theory, this does not prove anything, because social and cultural factors may mimic the action of natural selection regarding sex differences in behavior.[10]

Thus, one reason why it is impossible to separate biological from cultural determinants of human behavior is that either hypothesis can lead to similar predictions about how the behavior develops. Conversely, the fact that some observed behavioral trait can be shown to be adaptive should not be taken as evidence that it is biologically determined. For almost any human behavior one can successfully argue an adaptive, evolutionary, biological base *or* a socio-cultural one. Up to now it has proved impossible to devise or implement experiments that rigorously test either explanation. But when phenomena can be explained on either basis and neither explanation is testable, there are no scientific criteria for preferring either one.

Probably the most important point concerning the nature-nurture controversy is that to try to separate genetic and environmental components and discuss them separately is meaningless. Behavior results from the joint operation of genes and the environment, and these factors interact in complex and nonlinear ways that are different and unpredictable for different traits. In the absence of knowledge about these interactions for each specific trait one wishes to consider — and at present we do not have it about a single observed human behavior — the only meaningful question that can be asked is how much of the observed *variation* in behavior among individuals is caused by genetic factors and how much by environment. (And there are methodological difficulties even in answering this question.) This more limited question tells one *nothing* about how to partition genetic and environmental effects for the behavior itself;[11] nor does it tell us anything about the proportion of genetic and environmental contributions to the variance of any other trait.

What is more important to realize, and usually not acknowledged, is that even if one succeeded in partitioning the *variation* for any given trait, the answer would have *no* predictive value for different environments, because one cannot assume that the relative contributions of genetic and environmental factors remain the same as the environment changes. Indeed, there is experimental evidence that the relative importance of genetic factors for a given trait varies with the environment.[12] Thus, in comparing two groups that differ genetically, it is impossible to distinguish the genetic and environmental origins of *any* behavioral differences between them as long as their environments differ in *any* way. Moreover, even if one could study two groups in the same environment and overcome the problems of separating the variations caused by nature and nurture, one would still have no way to predict the effects of *any* changes in the environment. With reference to people, these methodological restrictions are equivalent to saying that questions about the relative contributions of genes and environment to differences in aptitudes or behavior between the sexes, races, or classes (or indeed any other groupings) *cannot* be tackled scientifically at this time.

Although it therefore is impossible to determine whether any human behavioral trait is genetically or environmentally determined, sociobiologists suggest that evolutionary reasoning can overcome this impasse. They argue that biologically determined behavioral traits can be identified by establishing their prevalence in many different societies. The underlying assumption is that if a given behavior were to be observed in all societies, this would prove it to be primarily genetic — the manifestation of a universal human biogram. In addition, if nonhuman primates were observed to behave in similar ways, sociobiologists claim that this strengthens the argument that the behavior has evolved through natural selection, and therefore is genetically programed.

There are difficulties with these arguments. The fact that individuals or groups have similar traits does not permit one to conclude that these traits share the same evolutionary and genetic basis. Biologists recognize this and use the term *analogy* to describe traits that look alike or function in similar ways, but that are not based on common descent. Analogous traits are examples of what is called convergent evolution: different evolutionary pathways that have come up with similar solutions to similar biological or ecological problems. Examples of analogous structures are the wings of bats, birds and insects, or the eyes of mammals, insects and squid. Although they look similar and subserve the same function — flight or vision — they have evolved separately and independently, hence are not related genetically. Traits that share a common evolutionary and genetic basis are called *homologous*. The commonality of their descent is often not obvious from superficial inspection, but usually is deduced from careful and systematic examinations of fossilized structures in the paleontological record. Examples of homologous structures are the scales and feathers of reptiles and birds. Though they look different, they may serve similar functions (covering, insulation) but most importantly, they can be shown to have the same evolutionary and genetic basis.

Clearly the study of behavioral traits is not amenable to this kind of analysis. Therefore it is impossible to *prove* that similar behaviors are something other than similar (that is, analogous) solutions to problems set by similar biological and environmental demands. Similarity of behavior is no proof of common evolutionary or genetic descent. But when it comes to sex differences, the situation is worse than that. Even if we take the example that is used most often to "prove" the genetic basis of behavioral differences, the suckling and nurturing behavior of human mothers toward their infants, a cool assessment of the facts shows two things:

(1) There is no evidence that the common styles of human mothering do not merely represent similar (that is, analogous) solutions to similar situational arrangements that juxtapose a mother whose breasts are painfully engorged with milk with a hungry and helpless infant. Clearly, milk, hunger, and helplessness are evolutionarily programed, biological factors. Need the nursing and nurturing also be genetically programed to explain what one observes?

(2) Among humans, the propensity for mothering by no means is a female universal.[13] Given various options, women respond quite differently to the challenge of childbearing. When a choice has been possible, some women have decided not to have children. When this choice has been blocked, some have given their children to various caretakers, and have done so even when it has been clear that this would mean almost certain death for the child (as was the case for many poor women in Europe as recently as the eighteenth and nineteenth centuries). Women who could afford it have hired wet-nurses and other substitute nurturers; some have left their infants with lactating relatives, neighbors, or friends. Women have used bottles and other gadgets that let older children, fathers and others nourish infants. Women even have exposed or otherwise killed their newborns or young infants.[14]

It is going to be difficult to establish how much of this has been choice, how much necessity. Nevertheless, there is good reason to suspect that the propensity of females for mothering may be a convenient, social fiction that ties women to reproduction and so frees men for production. (This is not to say that women are not an essential part of the productive work force today, just as they always have been. However, the myth of the innateness of mothering makes it easier to undervalue and underpay women, by assigning a marginal status to their contributions to production.) "Good mothering," one historian says, "is an invention of modernization."[13] And the message that we are given in the recent books by Bernard,[14] Lazarre,[15] Rich[16] and by Leibowitz[17] among others is that mothers' unique propensity for mothering is a fiction to boot.

The attempt to establish the genetic basis of a behavior by documenting its occurrence in all known human societies, encounters yet other difficulties. If a trait were shown to be universal and if we were to agree that it represented a genuine, unique line of evolution rather than the convergence of many evolutionary lines upon a common solution to the same or similar problems, we still could not say to what extent it had been genetically determined. Its very universality would indicate that the trait was insensitive both to genetic variability and to observed environmental differences. It would be fair to conclude only that the human genetic make-up is a *factor* in the appearance of the trait (as it is, of course, in the appearance of all traits), not that it determines it. We would have gained nothing of predictive value from the observation of a "universal" behavioral trait, since we cannot know what would happen in any as yet unobserved culture; the behavioral trait might or might not be expressed.

Whether or not universally observed behavior has anything to tell us about genetic causation is, however, only a theoretical problem. No behavioral traits have, in fact, been found that are common to all known cultures.[18] Nevertheless, sociobiologists try to make a case for the existence of universals, and for this they frequently rely on claims of the universality of sex differences in behavior among human beings. According to Barash:

> Sociobiology relies heavily upon the biology of male-female differences and upon the adaptive behavioral differences that have evolved accordingly. Ironically, mother nature appears to be a sexist . . .[19]

Sex differences are chosen, perhaps, because divisions of labor and of socially sanctioned behaviors by sex in fact do exist in almost all cultures. However, it is important to notice that the actual tasks and behaviors that are deemed appropriate for women and for men differ markedly from society to society.[18] [20] No differences in sex roles, behaviors or tasks are common to all societies other than the differences that are directly connected with the *biological* phenomena of reproduction and lactation. (Yes, wherever we look, only men inseminate and only women gestate and lactate.) Only the selective use of data enables sociobiologists to claim that there are universal sex differences in behavior; only false reasoning allows them to conclude from this that these differences therefore are genetically determined.[21]

The same fallacies haunt sociobiological interpretations of sex differences in behavior among nonhuman primates. Of course such differences exist. But among animals, just as in human societies, sex differences in behavior differ from group to group, even within the same species. For example, hamadryas baboons on open plains show aggression and male dominance, while in a forest environment (apparently less stressful) they show little sign of either.[22] In discussing conservative traits (those that are found in both nonhuman primates and in people) sociobiologists ignore this variability, claiming that primates show constant sex differences in behavior. Thus they misinterpret the fact that among all primates males and females exhibit some behavioral differences, and conclude that these differences are the *same*, and hence genetic (itself a fallacious deduction, as we have shown above).

Let us take Wilson's formulation of human sociobiological theory as representative. The traits Wilson considers to be "appropriate for hypothesis formation," and which he claims to be both universal and conservative, are the following:

> ... aggressive dominance systems with males generally dominant over females; scaling ... of responses, especially during aggressive interaction; intensive and prolonged maternal care, with a pronounced degree of socialization in the young; and matrilineal social organization.[23]

All these traits are based on sex *differences*, not on species universals, since Wilson argues that males are naturally more aggressive than females. Other social traits that develop from the basic ones listed are then claimed to be universals. An hierarchical structure to society follows directly from "aggressive dominance systems;" a sexual division of labor is derived from the aggressive dominance systems and from maternal care; economics and trade are said to have developed out of the resulting division of labor, through barter between females and males of hunted meat and gathered food; out of the development of trade came deception and hypocrisy. Territoriality, a lust for war and xenophobia, all alleged universals, are also results of aggression and of dominance systems.

Sexual selection (one of a number of possible forms of natural selection) is the evolutionary mechanism postulated as responsible for the development of the human species. Wilson quotes Robin Fox:

> ... sexual selection was the auxiliary motor that drove human evolution all the way to the *Homo* grade ... a premium would have been placed on sexual selection involving both epigamic [courtship] display toward females and intrasexual competition among the males.[24]

There are some interesting implications of this theory of the development of *Homo sapiens*. First, the development of sex differences, hierarchy and competition are seen as *intrinsic* aspects of human nature. Second, since only males compete, this theory leaves us with the familiar Victorian, Darwinian picture of passive females acting as carriers of genes selected through male action.[25]

Biosociology: Though sociobiology has existed as a formal entity only since 1975, criticisms of some of its many obvious scientific and logical flaws have begun to put it into the shade. For example, at a recent two-day symposium on sociobiology in Washington sponsored by the American Association for the Advancement of Science, many biologists and anthropologists disassociated themselves from this "new synthesis."

In the last year Alice Rossi, a feminist, has published two articles introducing her newer, "biosocial" science.[26] In the second of these articles, in *Human Nature*, Rossi criticizes sociobiologists for their "largely speculative leaps across the great differences that separate animal species." Unfortunately, she goes on and makes such leaps, but without acknowledging that that is what she is doing. Eschewing the particular evolutionary model of sociobiology, she says that in contrast "biosocial science searches either for critical contributions of physiological factors to human behavior or for the impact of social and psychological factors on body functioning." Hence she claims that the biosocial model is interactive, based only on humans, and that in it physiology and behavior are seen as interdependent. In reality Rossi attributes a determinative primacy to presumptive effects of "sex hormones" (and specifically of the "male" hormones or androgens) on the embryonic brain during a specific early stage of development. According to the model, these hormones "program" the brains of female and male embryos to react *differently* to the later demands of the spheres of reproductive or productive work, as defined in our society.

"The impact of gonadal hormones during fetal development," writes Rossi,

> now appears to play an important role in sex differences
> in adult behavior. The evidence is strong for critical

periods of hormone activity that affects the organization of the brain.

What she does not write is that this "strong evidence" has been obtained with experiments on laboratory rats and that the behavioral sex differences have to do with their specific reproductive functions, that is, with how male and female rats position themselves during coitus: the "mounting" behavior stereotypic for male rats and the back-bending or "lordosis" of the females.[27]

In the next sentence she continues:

> In the human fetus it is the presence of androgens, such as testosterone, that turns the fetus into a male. Without androgen, the fetus develops into a female.

Now, it is correct that not only humans, but all mammals develop as females, in the sense that they come to possess female reproductive structures (ovaries, uterus, and vagina), unless the primitive gonadal primordia are exposed to androgens at an early stage of fetal development. But this says nothing about the effects of androgens in programing the human brain. In support of that "largely speculative leap ... across species" Rossi cites observations and interpretations by John Money, Anke Ehrhardt and their co-workers. These investigators have adduced the best and clearest documentation of the overwhelming effects of early socialization on behavioral differences between boys and girls.[28] However, they also conclude that girls exposed prenatally to such abnormally high amounts of androgen that they are born with ambiguous or frankly male genitalia (which subsequently are corrected surgically to female genitalia), exhibit some behavioral differences when they are compared with normal girls, who serve as "controls." As Rossi put it, they "behave like tomboys despite their feminine upbringing."

Money and Ehrhardt's work on "tomboyism" has been criticized by many people.[29] Their definition of "tomboyism" is loose and culture bound, so that some of it was outdated a few years later. For example, how does one interpret today the datum that these girls "preferred more practical clothes" and "usually ... chose to wear slacks or shorts and skirts rather than dresses?" Or how does one ascertain that they are "lacking the enthusiasm of the control girls for becoming mothers of little babies," though they "did not reject the idea of having children?" What does one make of the observation that girls, who were born with penises that required surgical removal and many of whom received hormone treatments at regular intervals, "lagged behind their age mates in beginning their dating life and venturing into the beginnings of love play?" (Note, incidentally, that the oldest of these girls was 16½ when this description of her was written.) It is important to realize that none of these behaviors was observed by the investigators; they were all self-reported by the girls and their mothers.

The entire body of research on fetally androgenized girls starkly raises the question of how one provides scientific "controls" for children whose birth is hailed not with a cheerful "It's a boy!" or "It's a girl!" but with a halting, "I have something to tell you, Mrs. Jones. Your child is — I think — a girl; at least we think we can make her be one."

In addition to the work on sex hormones, as evidence for her theory, Rossi also cites behavioral studies which "suggest that women have many unlearned responses to infants." These studies consist of observations of maternal behavior (such things as the arm used to hold an infant, reactions to newborns, use of baby talk). As we have already pointed out, observations of common responses to infants need not imply that the behavior is genetically determined. Furthermore, such studies are useless for discussing possible sex differences in responses to infants (from whatever causes), because male responses under the same circumstances are unknown. Instead of evidence to support the assertion of sex differences in unlearned responses, we are offered speculations on genetic causation (for example, "It makes good evolutionary sense for mothers to have developed built-in positive responses to infants.") and an evocation of universality ("... but in all societies the mother has been the primary caretaker of her infants.")

Yet this kind of feeble evidence leads Rossi to conclude that though

> powerful cultural pressures may drown out physiological propensities ... the brain's predispositions affect the ease with which males and females learn socially defined appropriate behaviors for their gender

and to pretend that we have *evidence* that these "predispositions" are programed into the *human* brain by fetal androgens during a critical period of early development.

It would be foolish to argue that this programing may not happen; but on present evidence, it is not only foolish, but irresponsible, to assert that it does. Rather, we assume that people develop through multiple, ongoing interactions with their environments, which of course include other people, and that we, at present, have no way of sorting out these interactions. Such a continuously interactive model accounts best for what we know about the interplay of nature and nurture.[30]

In our opinion, the search for unique causes of behavioral sex differences is doomed, whether it sets the locus of causation in evolutionary prehistory, in the effect of sex hormones on the prenatal differentiation of the brain or in specific patterns of socialization. There is no evidence for Rossi's claim "that because of their evolutionary history, men and women differ in their predispositions to care for infants, and in their abilities to learn those caretaking skills;" nor do we know of any ways to obtain such evidence. Indeed, Rossi herself weakens this (sociobiologic) assertion with a "What I do believe is . . ." But in the very next sentence, she employs the authoritative voice of the scientist to state without qualification: "In comparison to the woman's attachment to her infant the man's attachment is socially learned."

Needless to say, this sentence means nothing. How does one measure women's or men's "attachment" in ways that permit "comparison?" And how would one go about determining the comparative inputs of "social learning" and "physiological propensities" to specifically sex-differentiated "attachments" of mothers and fathers to their infants? The degree of attachment of mothers to their young infants differs widely, as does that of fathers, *despite* the fact that girls usually are socialized to think of childbearing as their most important source of personal fulfillment and satisfaction. Yet this kind of value-laden language successfully masquerades as science when it is used authoritatively by persons with established scientific reputations. (A cynic might ask why, if parenting is all that natural for women, so many men have gained fame and fortune by telling us how to do it.)

The Social and Economic Questions

We have pointed out why neither sociobiology nor biosociology need be taken seriously as a scientific theory of human behavior. Unfortunately we do have to take them seriously as political theories. Theories of human behavior have the potential to shape behavior, for if people believe a certain characteristic to be genetically programed or innate, they are ready to believe that it can be changed only with difficulty or not at all. There is no better self-fulfilling prophecy than to ascribe a trait to "human nature." If women believe that they are innately more submissive than men, they become less likely to assert themselves. If men believe themselves incapable of parenting, they feel needlessly awkward when they first try. (What woman does not feel awkward when she first tries? Indeed, how many women feel defeated and depressed by how *unnatural* it feels to "mother" their first infant?)

If people believe that differences between groups are biologically determined, their belief is bound to affect the design of social programs to eliminate inequalities between groups. Support for the ERA, for example, has already been affected by a belief that a woman's place is with her family. "Science" now comes along to give this belief the status of natural law. Barash writes:

> ... women have almost universally found themselves relegated to the nursery while men derive their greatest satisfaction from their jobs...such differences in male-female attachment to family versus vocation could derive in part from hormonal differences between the sexes...[31]

And here is Rossi (leaping across species again):

> Rhesus monkeys and human males can become good fathers if they put their minds to the task or if they are forced into it by need or proximity; but it takes them a long time, full of trial and error.

(Rossi does not mention that not only human females but even Rhesus mothers apparently must learn how to mother.)[32]

Since the publication of E.O. Wilson's book, with wide publicity by Harvard University Press, sociobiology as a field has received much attention. *The New York Times* announced the theory to be a revolutionary breakthrough.[33] A flood of articles about the theory has appeared in such diverse publications as *Home and Garden, People, Time, Psychology Today, The New York Times Sunday Magazine,* The *Boston Globe* and The *National Observer.* A film has been made for high school use,[34] and a program of the Nova series on educational television was dedicated to sociobiology. Scholarly articles on sociobiology are common, and symposia, often led by prominent sociobiologists, have been scheduled at a variety of professional meetings, ranging from those of psychotherapists to sociologists and philosophers. Included in these is the above-mentioned two-day symposium on the subject at the annual meeting of the American Association for the Advancement of Science in February, 1978.

A second generation of books on sociobiology, more popularly written than Wilson's formidable work, have appeared, among them *The Selfish Gene* by R. Dawkins,[35] and Barash's *Sociobiology and Behavior.*[10] The theory has had an almost immediate effect on education. Courses on sociobiology are being given in many universities and colleges; material is being incorporated into high school and even grade school texts. A high school curriculum designed by sociobiologists R. Trivers and I. DeVore is used in over 100 school systems in 26 states.[36]

The newer biosociology is beginning to receive its share of attention. When Alice Rossi, respected sociologist with impeccable feminist credentials, asserts on the flimsiest of evidence the biological need for sex differentiated socialization, she is given a forum in a respectable scholarly journal and in a magazine designed for a more popular audience. Not surprisingly, her theory has been particularly noted by feminist scholars.

One must understand that these theories are not arising in a social and economic vacuum, but in a society that urgently needs them in order to stem and, if possible, reverse the *minimal* economic gains women (and minorities) have achieved in the present decade, largely as a result of political activism in the previous one.

According to United States Labor Department statistics released for May, 1978, 49.5 percent of adult (over age 20) women are in the labor force, and they account for 37.2 percent of it, while adult men account for 53.3 percent. The Department in March 1977 estimated that two-thirds of working women needed to work for financial reasons. In many occupations real wages have been decreasing and men's wages alone often no longer support a family. There are also a large number of self-supporting single women. On the other hand, women are important for industry. Women's pay is less than half that of men's on the average and "female" occupations have been expanding fasther than "male" occupations. Women in the labor force thus help keep labor costs down. However, given the stubbornly high level of unemployment, officially admitted to have been at or above 7 percent for more than a decade, the increase in the number of working women from 16.7 million in 1950 to 21.2 million in 1960 to 28.3 million at the beginning of this decade has raised complex questions. On the one hand, women have traditionally been the most expendable part of the labor force during times of high unemployment, and on the other, there are pressing economic reasons for keeping low-paid women in the labor force. In the last months, although unemployment for adult men has remained constant, unemployment for adult women has increased.[37] It is not clear whether this is a permanent trend. What is clear is that no matter what happens, the myth that women's place is in the home can be manipulated to serve an economic purpose by downgrading women's contributions to the labor force.

It is not a coincidence that in recent years a great deal of attention has been paid to the biological determinists who argue an updated version of "women's proper place." The political implications of these theories are clear and have been actively promoted by their originators and by others who stand to gain from them. The basic message is two-pronged: first, that important aspects of contemporary American society are shaped by biological factors and we must be wary about trying to change them; second, that attempts to change will fail or be expensive, since such efforts go against our natural propensities. David Barash longingly suggests: ". . . there should be a sweetness to life when it accords with the adaptive wisdom of evolution."[38] And Robert Trivers says that "It's time we started viewing ourselves as having biological, genetic and natural components to our behavior, and that we start setting up a physical and social world to match those tendencies."[39] The political message is explicit in Alice Rossi's urgings that the differing propensities of the sexes, such as women's biological suitability for child rearing, be taken into account when social policies are formulated. By trying to force answers to flawed questions about nature and nurture — questions that cannot be answered scientifically at the present time — sociobiological and biosocial theories end up simply as propaganda that pictures as inevitable the society in which the theorists have grown up and which has served them well.

A defense of the *status quo* through theories of biological determinism is not new. In the nineteenth century, widespread interest in similar determinist theories also coincided with a period of social unrest and questioning. It is in the interest of those with the greatest stake in keeping the society unchanged to promote such theories. And, of course, these people also are the ones who are in positions to influence opinion forming institutions, such as education and the media, and to advise the agencies that fund research.[40]

NOTES AND REFERENCES

[1] C. Dulles, "Vast Changes in Society Traced to the Rise of Working Women," *New York Times,* November 29, 1977; p.1.

[2] For discussions of the social and historical roots of these controversies see Allen Chase, *The Legacy of Malthus: The Social Costs of the New Scientific Racism* (New York: Alfred Knopf Co.), 1977, and Ann Arbor Science for the People, *Biology as a Social Weapon* (Minneapolis: Burgess), 1977.

[3] E. O. Wilson *Sociobiology, The New Synthesis* (Cambridge: Harvard University Press), 1975.

[4] A. S. Rossi, "A Biolsocial Perspective on Parenting," *Daedalus,* Vol. 106, 1977, pp.1-31; "The Biosocial Side of Parenthood," *Human Nature,* Vol.1, No.6, June, 1973, pp.72-79.

[5] In their papers in this book, Bleier and Salzman criticize the methodologies of these experiments as well as their interpretations — particularly the ready transfer of data collected with rats to the lives of women and men.

[6] E. O. Wilson, "Human Decency is Animal," *The New York Times Magazine,* October 12, 1975, pp.47-48.

[7] For a further discussion see: Sociobiology Study Group, "Sociobiology — A New Biological Determinism," in Ann Arbor Science for the People, *Biology as a Social Weapon* (Burgess), 1977, pp.133-49.

[8] E. O. Wilson, *Sociobiology, The New Synthesis,* (Harvard Univ. Press), 1975, p.159.

[9] J. Douglas and J. Miller, "Record Breaking Women," *Science News,* Vol.112, 1977, p.173.

[10] D. Barash, *Sociobiology and Behavior* (New York: Elsevier), 1977, p.283.

[11] R. C. Lewontin, "The Analysis of Variance and the Analysis of Causes," *Am. J. Human Genetics,* Vol.26, (1974), pp.400-411; "The Analysis of Variation and the Analysis of Causes," in N. Block and G. Dworkin, eds., *The IQ Controversy* (New York: Pantheon), 1976.

[12] W. Bodmer and L. Cavalli-Sforza, *Genetics, Evolution and Man* (San Francisco: W. H. Freeman), 1976, pp.435-465.

[13] For a recent review of this disturbing social reality, see Rosemary Dinnage, "Throwaways," *New York Review,* June 29, 1978, pp.37-39.

[14] J. Bernard, *The Future of Motherhood* (New York: Dial Press), 1974.

[15] J. Lazarre, *The Mother Knot* (New York: Dell Publishing Co.), 1976.

[16] A. Rich, *Of Woman Born* (New York: W. W. Norton), 1976.

[17] L.Leibowitz, *Females, Males, Families: A Biosocial Approach* (N. Scituate, MA: Duxbury Press), 1978.

[18] There is a vast anthropological literature documenting the range of cross cultural variation. See M. Sahlins, *The Use and Abuse of Biology* (Ann Arbor: University of Michigan Press), 1976, for a review of some of the evidence that important human social behavioral traits are not universal.

[19] D. Barash, p283., *op. cit.*

[20] For examples, see M. Mead's classic *Male and Female* (New York: Dell Publishing Co.), 1949; (reprinted 1968); R. Rorlich Leavitt, *Peacable Primates and Gentle People* (New York: Harper and Row), 1975; J. Nance, *The Gentle Tasaday* (New York: Harcourt Brace Jovanovich), 1975; C. Turnbull, *The Forest People* (New York: Simon and Shuster), 1962; R. Reiter, ed., *Toward an Anthropology of Women* (New York: Monthly Review Press), 1975.

[21] B. Chasin, "Sociobiology: A Sexist Synthesis," *Science for the People* Vol.9, No.3, 1977, p.30.

[22] D. Pilbeam, "The Naked Ape: An Idea We Could Live Without," in D. Hunter and P. Whitten, eds., *Anthropology: Contemporary Perspectives* (Boston: Little, Brown and Company), 1975, pp.66-75.

[23] E. O. Wilson, *Sociobiology, The New Synthesis*, (Harvard Univ. Press), 1975, p.551.

[24] *Ibid.*, p.569.

[25] See also, R. Hubbard, "Have Only Men Evolved?" in R. Hubbard, M.S. Henifin and B. Fried, eds., *Women Look at Biology Looking at Women* (Cambridge, MA: Schenkman Publishing Co.), 1979.

[26] The following quotations are from Rossi's recent article in *Human Nature*. Her ideas are laid out in the two references cited in footnote 4 above.

[27] For a discussion of the pitfalls that cannot be avoided when one tries to generalize from animal to human behavior, particularly with regard to these kinds of hormone studies, see Bleier's contribution to this volume.

[28] J. Money and A. Ehrhardt, *Man and Woman, Boy and Girl* (Baltimore: Johns Hopkins Press), 1972.

[29] See Salzman's article in this collection; also B. Fried, "Boys Will Be Boys Will Be Boys: The Language of Sex and Gender," in R. Hubbard, M. B. Henifin and B. Fried, eds., *op. cit.*

[30] See Salzman's article.

[31] D. Barash, p.301.

[32] H. F. Harlow and M. K. Harlow, "The Affectional Systems," in A. M. Schrier, H. F. Harlow and F. Stollnitz, eds., *Behavior of Nonhuman Primates*, Vol.2 (New York: Academic Press), 1965, pp.287-334.

[33] B. Rensberger, "Sociobiology: Updating Darwin on Behavior," *The New York Times*, May 28, 1975, p.1.

[34] *Sociobiology: Doing What Comes Naturally*. Distributed by Document Associates, Inc., 880 Third Avenue, New York, New York 10022.

[35] R. Dawkins, *The Selfish Gene* (New York: Oxford University Press), 1976.

[36] Educational Development Center, *Exploring Human Nature* (Cambridge: EDC), 1973.

[37] All these statistics are from an Associated Press story by Kristin Goff, "A New Look at Unemployment: More Women in US Labor Force," *Boston Globe*, June 5, 1978, p.12.

[38] D. Barash, p.310.

[39] From the soundtrack of reference 34.

[40] G. Domhoff, *Who Rules America* (Englewood Cliffs: Prentice-Hall), 1967; C. Karier, *Shaping the American Educational State: 1900 to the Present* (New York: Free Press), 1975.

SEX DIFFERENCES AND THE DICHOTOMIZATION
OF THE BRAIN: METHODS, LIMITS
AND PROBLEMS IN RESEARCH ON CONSCIOUSNESS*

Susan Leigh Star

Program in Human Development and Aging
University of California
745 Parnassus Avenue
San Francisco, California 04143

Introduction

The discovery that the two halves of the brain are specialized for different functions is being more eagerly explored by neurologists, psychologists, and other social scientists than perhaps any other single "fact" about consciousness and the brain. Literally thousands of articles have appeared in the last decade discussing, speculating, theorizing, documenting, and wishing about hemispheric asymmetry. A substantial portion of these articles deal with sex differences in this asymmetry, or lateralization. In sources ranging from the *Boston Herald* to *Neuropsychologia*, both scientists and journalists have hailed right and left brain differences as the "solution" to the enigma of sex differences — why men and women think differently, and appear to have different abilities.

It is my contention that nearly all the articles on sex differences in brain asymmetry, and a good many of those on other facets of the subject, are based on a network of interlocking assumptions which have no foundation in observed or observable reality, but which are sexist, political, reductionist and dangerous. They are epistemologically connected closely with other research in the area of sex differences which has called upon "biology," or upon what

* The writing of this paper was partially supported by a Public Health Service training grant from the Human Development and Aging Program, University of California, San Francisco, #AG00022-10 5T01.

writers agree to be biology, to "prove" things they would like to believe about women and men.

There are several different levels at which a critique is needed in order to understand this phenomenon, and I shall address them in turn: I shall begin with assumptions about sex differences research in general, examine the notions of localization of function and biological reductionism, move to what I perceive to be the underlying mythological (or pre-theoretical in Berger's sense)[1] bases, and finally, examine the data and actual research on hemispheric brain asymmetry and sex differences.

I. Sex Difference Research: Assumptions

Philosophers of science are careful to distinguish between *explanation* and *description*.[2] Writers about sex differences frequently confound the two, as well as making factual errors in both areas. While, of course, it is true that our attempts to explain are intertwined with our descriptions, and always influence the extent of what we are able to observe/describe, it is important for analysis to distinguish between the two. This is particularly important in areas that have a heavy emotional/political component, as is the case with sex difference research. For example, in the case of brain asymmetry research, some research observations are accurate, precise, and repeatable. But the explanations attached to them often ignore crucial explanatory variables, such as the subjects' training, experience, and education, and experimenter demand characteristics (expectations). On the other hand, some observations are simply unreliable or unfounded on any reliable empirical method; the explanations created from them are not worth testing.

Sex difference research, and particularly that on hemispheric asymmetry, has several competing and contradictory lines of thought in the areas both of explanation and of description. The section on mythology, below, deals with explanation; the methods section, with both descriptive errors and the explanations derived from them. The remainder of this section deals with general assumptions that affect both explanation and description.

The very fact of dividing subjects into male and female categories for research purposes may serve to reify and perpetuate a socially created dichotomy. The search for differences can help to create the differences; if you are looking for something you are likely to find it (cf. Rosenthal).[3] But the more basic problem is that the research on sex differences takes place within a gender-stratified society. That is, we are born into a sexual caste system and are heavily indoctrinated from birth with its values;[4] this system is stratified (as opposed to being simply differentiated) by virtue of the male-defined hegemony over money, power, and language. This fact presents practical problems that are not trivial in obtaining accurate descriptions. Given that sexual politics permeate all human interactions, including those within scientific laboratories, there is no way to "control" for this variable. A male experimenter in a white lab coat will elicit differential responses from male and female subjects because of prior socialization on their parts. Such implicit effects are not usually taken into account in sex difference research.

However, even if the level of description of the differences were accurate, given the impossibility of separating the nature-nurture interaction, we have no way of knowing that the genesis of the differences is "innate." If one finds a correlation between behavioral and biological differences, there is no way to know that the biology has produced the differences in behavior, rather than that the behavior has produced the biological correlate (for example, differences in arm and leg musculature of women and men, or differences in their verbal abilities).

One of the central ways in which sex difference research has been presented, and one of its key strengths for pro-patriarchal politics, is through confounding nature with nurture, and correlates with genesis: anything that can be found to have a biological correlate is interpreted as innate. Theorizers about sex differences go so far as to cite hormone research on rats and other animals to support their biological basis-behavioral genesis sequential reasoning. What they generally overlook is the fact that under most circumstances socialization, language, and environmental factors shape hormones, body size, and perception (including brain functioning) to a far greater and more reliably observable degree than the converse. Where biological *correlates* of behavior are reliably found, feminists need to ask if they are a reflection of socialization — and if they could be a useful way to examine socialization (and, by extension, to counteract negative or limiting socialization). One way of stating this question is to ask: Do we *somatize* our oppression? Rather than assuming that our bodies necessarily determine our social state, as patriarchal scientists have tried to prove, we must understand that social states can give rise to and shape many facets of our physical being. *Biology is no less, and perhaps in some areas, far more, mutable than socialization.*[5]

A further implication of the above is that biology and socialization are not really different from one another, or rather, that they coexist in an inseparable dialectic. We have no language at present that does not reflect a Cartesian nature/nurture dichotomy for discussing sex differences. It is difficult to resist the urge to ask, "But what, *underneath it all,* really *are* the differences between men and women?" *What we must begin to give voice to as scientists and feminists is that there is no such thing, or place, as underneath it all.* Literally, empirically, physiologically, anatomically, neurologically . . . the only accurate locus for research about us who speak to each other is the changing, moving, complex web of our interactions, in light of the language, power structures, natural environments (internal and external), and beliefs that weave it in time.

II. Brain Asymmetry: Mythology and Background

In human beings the brain is divided into right and left halves, which are connected but appear to be specialized for different functions. Broadly speaking, the left hemisphere is specialized to process linear, sequential, and "propositional" thought, the right hemisphere for simultaneous, Gestalt, and "appositional" thought.[6] This specialization is called hemispheric asymmetry, or cerebral lateralization.

The implications of these differences have often been embodied and popularized in theory which grounds itself in certain aspects of Eastern religions. Buddhism, for example, postulates two complementary types of energy, *yin* and *yang*, present in all life forms, which could be seen as roughly analogous to some of the left-right properties of the brain. *Yang* is seen as constricted, male, and propositional; *yin*, as expansive, simultaneous, and female. Some of the writing about brain asymmetry has not bothered to weed out these types of sexist stereotypes about "masculine" and "feminine" in its discussion of brain asymmetry. Robert Ornstein, a major figure in both the research on brain asymmetry and its popularization, draws upon old Buddhist doctrine to emphasize his points about left brain and right brain functions. For example:

> The Chinese Yin-Yang symbol neatly encapsulates the duality and complementarity of these two poles of consciousness ... Note that one pole is in time, the other in space; one is light, one dark; one active, one receptive; one male, one female.[7]

In doing this, he reifies and further extends common traditional stereotypes about masculine and feminine into the literature on brain asymmetry.

In the years since Ornstein's book was written, many psychophysiologists have revised their initial conceptions of brain asymmetry, and most, including Ornstein, now realize that there is no duality of consciousness in the brain, nor are the "energies" described above limited to men or to women. However, such realizations are rarely put forth clearly or forcefully in the popular or less-than-technical press, and even careful researchers in brain asymmetry still slip into the culturally-condoned and readily available language of sex differentiation to describe brain functions. The mythologies upon which they draw are not only esoteric or Eastern. Western myths as well have attributed different functions to the right and left sides of the body. The left side is named dark, mysterious, sinister and feminine; the right side is seen as light, logical, and masculine, corresponding to the yin-yang dualism of the East. For example, when the Osgood Semantic Differential Test was administered to a group of college students in America, they came up with the following results:

> The Left was characterized as bad, dark, profane, female, unclean, night, west, curved, limp, homosexual, weak, mysterious, low, ugly, black, incorrect, and death, while the Right meant just the opposite — good, light, sacred, male, clean, day, east, straight, erect, heterosexual, strong, commonplace, high, beautiful, white, correct ...[8]

The equation of the female with the left side of the body, and thus the right side of the brain, is a confusing mythological underpinning from the point of view of subsequent research, which began to identify the left cerebral hemisphere with verbal ability, at which females seem to be better than males. None of the research I have read has directly confronted this contradiction; rather, gender-marked words, such as "masculine" have permeated the field and contributed unnecessary confusion to it.

The dualism of Western language has added its own ironic dimension for feminist interpreters of popular psychology, some of whom have perceived the "left brain" as patriarchal, the source of too much linearity, and of non-holistic structures. Both feminists and those involved with Eastern culture (meditation, etc.) have called for a "return to the right brain." The underlying reasons for discontent are real, but the "grafting of philosophy onto anatomy" as Nebes[9] puts it, is inaccurate. The reasons why it is so are complex, and require some background history of the development of research and theory on brain asymmetry.

The current strong wave of interest in hemispheric asymmetry began with the work of Roger Sperry, who studied persons with severe epilepsy who had had their *corpora callosa* (the nerves and other tissue connecting the two halves of the brain) surgically severed in an attempt to control seizures. As a result, these people had two separately functioning brain systems — their right hand literally did not know what their left was doing (unless they had a chance to look and see!). Sperry presented the two sides of the brain in such people with a variety of stimuli — verbal, tactile, and visual. On the basis of his subjects' responses, he was able to generalize about the types of functions that the two hemispheres perform separately. He concluded that the left half of the brain determines logical thought, most speech, mathematical ability and "executive" decisions, while the right half rules visuo-spatial ability, emotions, and intuitions. His hypothesis about the left hemisphere controlling verbal functions was supported by observations on brain-damaged people with lesions in their left hemispheres, many of whom have more and severer speech impairments than those with right hemisphere lesions.

Sperry did find a clear-cut dichotomy in these surgical patients, but he and later investigators recognized that in normal, uninjured individuals the two hemispheres are interdependent[10] and that specialization was much more subtle than is suggested by the crude simple dichotomization into "verbal" and "spatial" that was to mark the literature for the next decade.

The strict equation of spatial ability with right hemisphere functions, and of verbal ability with the left, combined with the tenet that it is most natural to use one side of the brain at a time, formed the basis for subsequent theories about sex differences in asymmetry. Sperry's hypothesis, based on his observations of fewer than twenty patients,[11] became a "fact" that was subsequently used to build new theories.

Despite the demonstrated *interplay* between the left and right halves of the brain in the literature, verbal and spatial abilities had come to be so firmly identified with left and right hemispheres that many experimenters (and philosophers) simply equated superior verbal ability with "left brain dominance" and superior spatial ability with "right brain dominance," without recording EEGs (brain waves) or performing any other physiological *measurements* to determine which half in fact was more active during the specific tasks. Many of the major hypotheses about sex differences in hemispheric asymmetry are inferred from differences in performance on specific verbal and spatial tasks, and are based on little or no direct physiological measurement.

Thus, many of the hypotheses which form the basis for theory about sex differences are second-level, or second-hand, hypotheses that are not grounded in empirically observed, or, sometimes, even observable phenomena, but are inferred from differences in verbal and spatial abilities, as measured on tests ranging from college board scores to various "spatial tasks."

Several questionable assumptions underlie conclusions based on such methods. First, we cannot be sure that the tests measure anything more than one's ability to perform on these tests. It is an unfounded extrapolation to assume that they measure "verbal or spatial ability." This is especially true of tests such as the SAT verbal test, or the rod and frame test for spatial ability.

We also do not know that the "verbal or spatial ability" that is measured by these tests is the same as what Sperry (and later, others) deduced from his epileptic, split-brain patients. Nor can we be sure that all "spatial tasks" activate the right hemisphere and all "verbal tasks" the left hemisphere. Perhaps most importantly, we do not know that differences in performance on tests are due to differences in brain lateralization and not to prior training or experience.

Recent work on hemispheric asymmetry has stressed the importance of *individual* differences and the problems and complications that arise when trying to assess differences between *groups* of individuals, whether they be men and women, or artists and lawyers.[12] [13] [14] But these complexities are often disregarded in the search for sex differences.

Two hypotheses have been instrumental in the expansion of the literature on lateralization and brain asymmetry; both involve the sorts of methodological difficulties I have just described.

III. Two Competing Hypotheses about Right Brain/Left Brain Functions

The following two hypotheses have been among the most widely discussed and believed theories about sex differences in brain asymmetry. They are both based on alleged sex differences on spatial and verbal tasks, and represent two entirely different reasonings from basically the same set of "facts;" Levy and Sperry say that women are inferior on spatial tasks because of a *lesser degree of lateralization;* Buffery and Gray say that they are superior on verbal tasks because women's brains are *more lateralized.*[15]

The Levy-Sperry Hypothesis

Levy and Sperry begin their reasoning by noting that females perform poorly on certain tests for spatial abilities, and that left-handed men perform poorly on the same tests. Left-handers, they state, perform poorly on these tests because of "cross-talk" from their left hemispheres while performing the tasks: they are said to be less lateralized. The authors argue that the superiority of right-handed males in such spatial tasks is due to a *greater* lateralization of the brain: Levy states that "it might be that female brains are similar to those of left-handers in having less hemispheric specialization than male right-hander's brains."[16] She and Sperry also draw a further analogy between females and left-handers: they state that in left-handers language is mediated by both sides of the brain (whereas in right-handers it is a left brain function) and that the language component in the right hemisphere of the left-handers (alleged to be absent in right handers for the most part) is what interferes with "pure" right-hemisphere performance on spatial tasks. (In fact, it is *not* true that left-handers usually have bilateral language representation![7]) From this Levy and Sperry generalize to females who, they assume, also have bilateral representation for language, and they conclude that this is why females as a group perform more poorly than males on spatial tasks.

A number of researchers have already begun to accept their hypothesis as fact, and are using it to interpret further findings, although the problems with it are legion. Levy and Sperry do not address training and socialization as possible factors in performance of spatial tasks. They do not verify their assumption that the tests measure the degree of hemispheric specialization. And they do not address the critical fact that females consistently perform better than males on tests of *verbal* ability, a fact which would seem to contradict their assumption that females have bilateral language representation (which, by their reasoning, should make their verbal abilities *poorer*). Rather, they seem more interested in explaining male superiority on spatial tasks, whatever contortions of logic this might demand.

The Buffery-Gray Hypothesis

Buffery and Gray examine the same test scores as Levy and Sperry, which show that males perform better at certain spatial tasks; but unlike Levy and Sperry, they also take into account female verbal superiority.

To explain how both apparent superiorities can co-exist, Buffery and Gray construct the following hypothesis. They postulate that in males, linguistic and visuo-spatial abilities are represented in both hemispheres, whereas in females they are separated into the left and right hemispheres respectively. (Thus for Buffery and Gray, females are *more* lateralized than males, exactly opposite to Levy and Sperry's conclusion.) Buffery and Gray then assert that bilateral representation is most efficient for visuo-spatial tasks — a direct contradiction of most theories — because these tasks require a global, holistic perception. Hence males, with less lateralization than females, perform better on visuo-spatial tests. Then, with a confounding leap in logic, they assert that verbal tasks "require *more* lateralization," since they are more "specific" and "delicate" and "localized" than spatial tasks. Hence women, with greater lateralization than men, perform better at verbal tasks.

There are at least three serious problems with their hypothesis. The first is that deriving a more global or Gestalt perception from superior performance on the spatial tasks that have been used in these tests requires a bit of imagination. For instance, one task is the rod and frame test, which gauges the ability, in a darkened room, to adjust a movable glowing rod within a tilted frame to a vertical position. Another tests the ability to distinguish pictures of familiar objects that are concealed within a camouflaging background. The ability to take a figure out of its background context is called "field independence" and is used as an example of "spatial ability."

In these sorts of tests females, on the average, are less able to separate a figure from its context, and are therefore said to be more field dependent than males. From this it would appear that *females* are the ones who exhibit Gestalt perception (right hemisphere), yet this is attributed to *men* in the attempt to explain their supposedly superior spatial ability.

But a more blatant contradiction emerges from the Buffery-Gray theorizing. They casually mention that

> male superiority on visual tasks only appears when manipulation of spatial relationships is involved. On tasks which depend for their execution principally on the discrimination and/or comparison of fine visual detail, the direction of the sex differences is reversed. Thus women are better than men on . . . a *number* of other tests of visual matching and visual search. . .[emphasis mine][15]

Thus, the *only* tasks that show men are more able are tests of manipulation of the environment or some part of it. The equation of this with spatial ability, not to mention its high valuing, reflects the respect accorded male skills in this society.

Buffery and Gray end the above quote with: "Thus women are better than men on ... a number of other tests of visual matching and visual search which are predictive of good performance on clerical tasks."[18]

Finally, Buffery and Gray, like Levy and Sperry, never identify by means of physiological tests the hemisphere whose presumed activity they associate with a particular task. They *postulate* that men are less lateralized than women; they *postulate* that verbal skills require greater lateralization, and visuo-spatial skills less lateralization. But they never *measure* the brain activity of males or females during the performance of any of these tasks.

IV. Contradictions, Conclusions and Feminist Analysis

EEG Studies on Sex Differences in Brain Asymmetry

Recent experiments using the EEG (electroencephalogram — a recording of electrical activity within the brain, or "brain waves") have produced evidence that contradicts the Levy-Sperry and Buffery-Gray hypotheses. While these experiments suffer from the difficulties I mentioned in Section I, as does all work on sex differences, they show that the analysis must proceed from a more complex base than the above generalizations.

Davidson and Schwartz[19] and Davidson et al.[20] have recorded the EEGs of men and women during the performance of various tasks. They find that right-handed females show greater brain asymmetry than males on self-generated tasks and can also control the *amount* of asymmetry more precisely. For example, when asked to perform a "right brain" task, such as whistling, women's right hemispheres are more active and their left less so than are men's. This is also true for "left brain" tasks. When asked (after prior training) to produce more or less asymmetry in either direction, females are much better at both tasks.

Another task was to produce at different times, either an internal state of high emotion or a non-emotional state. This was done by either reliving a situation of intense anger, or by thinking of some nondescript topic. During the emotional tasks, females showed more right-hemisphere activity than males; during the non-emotional tasks, they showed less. Males showed no shift in asymmetry between the two emotional conditions. So, females showed greater asymmetry and greater differences in asymmetry between emotional tasks and non-emotional tasks than did males in this study.

The third task involved biofeedback training to enable people to control the amount of hemispheric asymmetry they were producing and to produce asymmetry at will. Females were better at producing greater asymmetry, even to the degree of using *only* one side of their brains. There were no sex differences when people were asked to use both sides of the brain simultaneously.

Davidson et al. generate the following sets of hypotheses on the basis of their work:

(1) They characterize the "male" way of feeling as much more analytical, more "left brain," whereas females may typically process emotions in a more global and Gestalt-like manner. In other words, during these experiments the male subjects were unable to produce a right-brain emotional state without left-brain interference.

(2) When asked to think about something without emotion, males were less able to do so than females. This provides an interesting twist to the traditional stereotype that women's "emotional" way of thinking clouds their rationality.

(3) Females have better control over the direction of their EEG asymmetry than do males — i.e., they can utilize *either* hemisphere more precisely depending on appropriateness.

Feminist Perspectives on the Research and Possibilities

What seems clear from the studies I have examined is that the imputation of male superiority on "visuo-spatial" tasks has been used as the basis for much of the theorizing about brain asymmetry. It is an artifact-laden generalization. Most of the test results can probably be attributed to training or socialization, and do not necessarily reflect inborn differences in brain functioning. (There is, furthermore, a question as to whether the scores themselves are accurate, given some of the situational variables mentioned above.) Even if men's superiority on these tasks were proven beyond a doubt, the real-life significance of the tested "visuo-spatial" tasks is questionable. What these researchers imply is that on tests of spatial skill, men are better at manipulating the environment, except when "focused, delicate" spatial skills are required. Manipulation of the environment is not necessarily a desirable skill, nor is the ability to take things out of context (field independence).

One thing that emerges from the data is that men seem to have difficulty employing their right cerebral hemispheres in a focused way, or to perform any but manipulative spatial tasks. Yet researchers have interpreted the test results to mean that men are superior in "spatial ability."

Two contradictory stereotypes emerge: men are linked with the right hemisphere because of their supposed superior spatial abilities, women with the left because of superior verbal scores. Yet, men are also thought to be "more analytic" (or "logical"), which would be expected to be linked with the left hemisphere. Complex and circuitous arguments have been required to come up with men's "superior spatial ability" while leaving the myth of their razor-sharp intellects intact.

Hemispheric asymmetry research at present is only a small part of the brain research that is being done on sex differences. But it forms an important basis for much of the theorizing in the area, and is also an exemplar of the pitfalls of skipping levels of analysis in psychological research.

Feminists utilizing the research on hemispheric asymmetry should be careful to avoid similar pitfalls and not take right and left brain dichotomies literally when analyzing women's oppression, or at least not use them in the simple literal fashion in which they are popularly represented. There *is* a strong metaphoric relationship between the popular dichotomizations for "right brain" and "left brain" functions/perceptions and some of the ideological/material differences between feminists and sexists. Some feminist values that have recently been articulated[21] include holistic perceptions, non-dualistic thought, and a validation of intuition. Patriarchal values are associated with linear thought, propositionality that "objectifies," and dualism. However, the linkage of male dominance on a social level with "the left brain" is too simplistic.

An example of this kind of oversimplification occurs in an article by gina:

So dualism resides in the very brain. The ways of perceiving that came to be grouped in the left hemisphere are the tools men used to take control of the planet. Linear thinking, focused narrowly enough to squeeze out human or emotional considerations, enabled men to kill ... with free consciences. Propositional thinking enables men to ignore the principles of morality inherent in all the earth's systems, and to set up instead their own version of right and wrong which they could believe as long as its logic was internally consistent ... All ways of perceiving that threatened the logical ways with other realities were grouped together on the other (right) side of the brain and labeled "bad."

The separation of "good" and "bad" qualities into left and right sides of the brain, and the universally constant valuation of qualities, can be seen in every patriarchal culture through its attitudes toward left and right-handedness ...[22]

Gina here introduces a dualism that *rejects* as male our ability to use the tools of intellectual reasoning and logic; and this, too, is dangerous for it perpetuates stereotypic masculine/feminine dualities, and even more subtly so if they occur in the same person. Our left hemispheres are not precarious, "male" places to be visited but not dwelt in. We need to utilize both halves of our brains in a flexible and adaptive manner, based, as gina suggests, in a moral society which respects the activities of both.

For feminists, our central concerns must be to eliminate patriarchal mechanisms that have blocked the expression and *validation* of language and spatial/intuitive/environmental skills in women, and to encourage the development of these skills in the holistic manner of which we seem to be capable.

The existence of a social system which encourages competitiveness between different modes of thought, and often emphasizes linearity at the expense of holism, does not imply that this system must derive from brain structure; nor need we assume that the system that exists supports or interacts smoothly with extant brain structures. The brain is so much more overwhelmingly labile than static, individual than standard, and state-dependent rather than constant over time, that we must find language that reflects this when talking about brain-society interactions — whether on the level of metaphor or actual function. For feminists, I suggest that we take the *corpus callosum* as our metaphorical and functional ideal locus — the always-changing, time-independent, inter-hemispheric conductive tissue, which connects the two hemispheres and hence cannot be separated analytically or physically from either.

NOTES AND REFERENCES

[1] Berger, P., *The Sacred Canopy.* Garden City. N.Y.: Anchor Books, 1967.

[2] I am grateful to Sarah Hoagland for pointing out this distinction to me.

[3] Rosenthal, R., *Experimenter Effects in Behavioral Research.* N.Y.: Appleton-Century-Crofts, 1966.

[4] Daly, M., *Beyond God the Father: Towards a Philosophy of Women's Liberation.* Boston: Beacon Press, 1973.

[5] Schacter, S., and Singer, J., "Cognitive, Social and Physiological Determinants of Emotional State," *Psychological Review, 69,* 379-399, 1962.

[6] Bogen, J.E., "The Other Side of the Brain II: An Appositional Mind," *Bulletin of the Los Angeles Neurological Society, 34,* No.3, July, 1969, 135-162.

[7] Ornstein, R., *The Psychology of Consciousness,* San Francisco. W.H. Freeman, 1973, pp.65-66.

[8] Domhoff, W., "But Why Did They Sit on the King's Right in the First Place?" in *The Nature of Human Consciousness,* Ornstein, R., ed., San Francisco: W.H. Freeman, 1973.

[9] Nebes, R., "Man's So-called Minor Hemisphere," *UCLA Educator, 17,* No.2, Spring, 1975.

[10] The two hemispheres are also interdependent *within* split brain individuals, since special screens which separate the two eyes and hands from seeing or touching each other are required in order to test patients. Presumably information travels between the two hemispheres via lower brain connections, or muscle feedback in other parts of the body. Levy, J., Trevarthen, C. and Sperry, R.W., "Perception of Bilateral Chimeric Figures Following Hemispheric Deconnexion," *Brain, 95,* 61-77, 1972.

[11] Sperry, R.W., "Lateral Specialization in the Surgically Separated Hemispheres," in Schmitt, F.O. and R.T. Wardon, *The Neurosciences: Third Study Program.* Cambridge: MIT Press, 1974.

[12] Kocel, K., Galin, D., Ornstein, R. and Merrin, E.L., "Lateral Eye Movement and Cognitive Mode," *Psychometric Science, 27,* No.4, 223-224, 1972.

[13] Galin, D. and Ornstein, R., "Individual Differences in Cognitive Style I. Reflective Eye Movements," *Neuropsychologia, 12,* 367-376, 1974.

[14] Galin, D., "Methodological Problems and Opportunities in EEG Studies of Lateral Specialization," to appear in *Symposium on Neurological Bases of Language Disorders in Children; Methods and Directions for Research,* National Institute for Neurological and Communicative Disorders and Stroke, 1978.

[15] Buffery, W. and Gray, J., "Sex Differences in the Development of Spatial and Linguistic Skills," in C. Ounsted and D.C. Taylor, eds., *Gender Differences: Their Ontogeny and Significance*. Edinburgh: Churchill Livingstone, 1972.

[16] Levy, J., "Lateral Specialization of the Human Brain: Behavioral Manifestations and Possible Evolutionary Basis," in Kiger, J.A., ed., *The Biology of Behavior*. Corvallis, Oregon: Oregon State University Press, 1972.

[17] Marshall, J., "Some Problems and Paradoxes Associated with Recent Accounts of Hemispheric Specialization," *Neuropsychologia, 11*, 463-470, 1973.

[18] The use of the word predictive here is interesting. It is an adjective with a deleted agent. The unasked questions implicit in the word are *who* predicts, for *whom* and *why*. For a linguistic analysis of "scientific" (or "scholarly") terminology like this which deletes the answers to the "who" and "why" questions, see Julia P. Stanley, "Syntactic Exploitation: Passive Adjectives in English," paper presented to the Southeastern Conference on Linguistics, Athens, Georgia, April 1972; and "Passive Motivation," *Foundations of Language, 13*, 25-39, 1975.

[19] Davidson, R.J. and Schwartz, G.E., "Patterns of Cerebral Lateralization During Cardiac Biofeedback Versus the Self-regulation of Emotion: Sex Differences," *Psychophysiology, 13* 62-68, 1976.

[20] Davidson, R.J., Schwartz, G.E., Pugash, E. and Bromfield, E., "Sex Differences in Patterns of EEG Asymmetry," *Biological Psychology, 4*, 119-138, 1976.

[21] Star, S.L., "The Politics of Wholeness II: Lesbian Feminism as an Altered State of Consciousness," *Sinister Wisdom, 2*, No.5, Spring 1978.

[22] gina. "Rosy-Rightbrain's Exorcism/Invocation," *Amazon Quarterly, 2*, No.4, 1974.

TRANSSEXUALISM: AN ISSUE OF SEX-ROLE STEREOTYPING

Janice G. Raymond

Hampshire College
Amherst, Massachusetts

The issue of transsexualism is thought to affect only a small percentage of the population, i.e., those few individuals who feel they are men or women "trapped in the body of the wrong sex." This is hardly the case, however. Transsexualism is an issue which raises important questions for all of us concerned about *why* it is in this society, that one could even talk about having, for example, a "female mind in a male body." Ultimately, transsexualism offers a unique perspective on gender identity, sex-role stereotyping, and sex differences in a patriarchal society, because it provides a graphic locus for issues which are less obvious in society-at-large. Put simply, here we have the stereotypes of masculinity and femininity on stage, so to speak, for all to see in an *alien* body.

Transsexualism is also an important medical ethical issue which raises questions of bodily mutilation and integrity, "nature" vs. technology, medical research priorities, unnecessary surgery, and the inevitable issue of the medical model in general.[1] It also raises questions about definitions of maleness and femaleness, and their boundaries. Feminists who are concerned about the takeover of female reproductive capacity inherent in genetic technologies (e.g., test tube fertilization and cloning), may find that similar questions about the control of female biology rear themselves in the trans-sexual context.

Scholars will also find that transsexualism touches the boundaries of many of the existing academic disciplines in such a way as to raise fundamental questions about the territorial imperatives of biology, psychology, medicine, and the law, to name but a few. Questions about the causes of transsexualism and the proper methods of treatment have been hitherto restricted to the domain of psychology and medicine. But as an ethicist and feminist scholar, I would maintain that these issues of causation and treatment are often suffused with male-defined values and philosophical beliefs — beliefs about the so-called natures of women and men, for example. When John Money, as one specific example, states that the core of one's gender identity is fixed by the age of 18 months, his statement is fraught with certain normative beliefs about individuals' abilities to change. Such beliefs can become invisible in the mounds of supposed scientific data about sex differences that Money offers.[2] Thus science often becomes metaphysics.

Gender dissatisfaction/agony is a very real problem and strikes at the heart of the transsexual issue. I am concerned with several major aspects of this problem and its effects. It affects not only transsexuals but also the wider community of those who suffer from the same gender dissatisfaction and who are deeply concerned about sexism, sex-role stereotyping, and sex differences. In this context, one of the major questions to be asked is what causes transsexualism — biology, psychological make-up, or family conditioning?

Biological factors are sophisticatedly invoked in the work of John Money and his associates. Their work has become a kind of bible on sex differences and has also enjoyed a wide acceptance among feminists.[3] Money is no biologizer of the *ancien regime* in which, for example, hormonal determinists linked anatomy directly with destiny. Rather, Money is careful to claim only that hormones "set the direction but not the extent" of sex differences, and that hormonal influences are mediated through the hypothalamus. In the case of transsexuals, something allegedly goes wrong during the prenatal critical period when sex hormones are taking effect.

Money's use of a mediating factor and his emphasis on socialization as completing the "program" of general sexual development distracts from his view of *biology as incipient destiny* and focuses attention more on the environmental side of the coin. In stressing the latter, Money claims to unite biological and environmental factors into a unique gestalt, telling us that the nature-nurture debate is obsolete. But few critics have noted that the theme of "biology is destiny" changes to "socialization is destiny." Money emphasizes that the socialization aspect of the "sexual script" is more significant than the biological. It is so crucial and happens so early that "core" gender identity is fixed by the age of 18 months. "When the gender identity gate closed behind you, it locked tight. You knew in the very core of your consciousness that you were male or female. Nothing short of disaster could ever shake that conviction."[4] Socialization takes on all the force of a new natural law. While few would deny that socialization is a very powerful factor, to give it Money's force and power is to make it absolute and static.

In summary, Money credits biology with too much destiny. Because he awards too much undemonstrated significance to prenatal, critical hormonal factors, he can be termed a biologizer. But also because he asserts that socialization is the new wave of destiny, he can be called an environmental determinist. A unique combination! In my opinion, he is wrong about both sides of the developmental coin. Thus I call him a pseudoorganicist who achieves only the appearance of reconciling nature and nurture. Given this state of affairs he tells us very little about the origins of transsexualism.

Not all transsexual theorists would emphasize hormonal factors or, for that matter, biological causes of any sort. Some see the cause of transsexualism in the intrapsychic and/or interpersonal background of the transsexual. Of these psychological proponents, Robert Stoller is perhaps the best known.[5]

Relying largely on clinical data from case studies and interviews with transsexuals and their families, Stoller is fond of highlighting what I call the "mother-smothering" factor. He attributes male-to-constructed female transsexualism to a reductionistic, symbiotic mother-son relationship which usually occurs within the context of a disturbed marriage. In the clinical cases he relates, Stoller claims that the mothers suffer from marked cases of classic Freudian penis envy. The boy child, later to become a transsexual, serves the mother as the treasured phallus for which she supposedly yearns. The child thus becomes "mother's feminized phallus" and does not pass through what Stoller accepts as the "normal" male formative stages of separation from the mother and consequent individuation. Thus he does not develop a "normal" masculine gender identity and has no fear of castration. The fathers of transsexuals, in Stoller's cases, were usually passive men who were consistently absent from their homes or, if physically present, were absent emotionally. Given this situation, each mother turned to her son for continuing fulfillment, concentrating totally on him and encouraging the boy to identify with her and with her body by constant physical contact. This symbiosis often lasted into pre-pubertal years, and the boy thus began to feel himself a female, despite the evidence of his senses that he was anatomically male.

Although the literature on transsexualism is overwhelmingly oriented toward the male-to-constructed female transsexual, 10 female-to-constructed male transsexuals have been studied by Stoller. Here he again focuses blame on the mother. The female transsexual who, in infancy, lacks "feminine graces" and is not "cuddly" supposedly has a mother unable to show any emotional tone due to frequently recurring states of depression. Again the father is passive and has little or no emotional rapport with his wife and child. As a result, the girl is used by both parents as a father-substitute to alleviate depression. Her acting-out of masculine characteristics is encouraged by both parents and becomes self-perpetuating.

It is my contention that as long as causation theories about transsexualism continue to measure a transsexual's adjustment or non-adjustment to the norms of masculinity or femininity, then they miss the point completely. I suggest that transsexualism, although it may be secondarily influenced by psychological and/or family conditioning factors, is primarily caused by the rigid sex-role stereotypes that a sexist society generates. Thus transsexualism is fundamentally a social and ethical issue.

In this view, the gender dissatisfaction that is expressed by transsexuals is symptomatic of something that is wrong on a deeper level. In this view, persons who desire to change sex surgically are fundamentally dissatisfied with the gender identity and role of their native-born sex to the point that the body and organs of the opposite sex come to incarnate the "essence" of their desired masculinity or femininity. Thus sex conversion surgery enables transsexuals to exchange one stereotype for the other, thereby reinforcing the fabric by which a role-defined society is held together. Only within a society in which rigid sex-role stereotypes exist does it make sense to surgically adjust a person's body to his or her mind (transsexual surgery), if the person's mind cannot be adjusted to his or her body (e.g., through traditional therapy).

It is significant that the majority of transsexuals are male-to-constructed females. There are many reasons for this. Most importantly, I think, men are socialized to fetishize and objectify. The same socialization that enables men to objectify women in rape, pornography, and "drag" is also that which enables them to take distance from their own bodies. Thus the penis is seen as a "thing" to be gotten rid of. The female body parts, specifically the female genitalia, are seen as "things" to be acquired. In the case of transsexual surgery, the female genitalia are viewed not only as parts of a female body which can be acquired at the hands of a surgeon, but also they are completely separated from the biological woman. At the surgical point, they come to be totally dominated by incorporation into the biological man. Medicalized transsexualism can thus be seen to be the ultimate conclusion of the male possession of women in a patriarchal society.

Medicalized transsexualism also creates male-to-constructed females who are more feminine than most feminine biological women. This highly stereotyped femininity is revealed in the way these transsexuals act, speak, and define themselves as women. One of the questions I asked transsexuals whom I interviewed was why they wanted to become women. Most responded in terms of the classic feminine stereotype — they wanted to be nurturing, passive, housewives, etc. Some expressed their decision to change sex in terms of desired marriage and motherhood — wanting to "meet the right man," "have him take care of me," "adopt kids and bring them up."

Very significant is the role of many so-called gender identity clinics and of the medical establishment in fostering and reinforcing stereotyped behavior. It is a primary requirement of these centers, in counseling persons who wish to be transsexed, that they must "pass" as "true women" in order to qualify for treatment and eventually surgery.[6] "Passing" requires evaluation of everything from feminine dress, to feminine body language, to feminine positions in intercourse. At Johns Hopkins, candidates for surgery are required to live out opposite-sex roles for at lease six months. Many other centers require a longer period, some up to two years. Thus the role of such clinics and clinicians in reinforcing sex-role stereotypes is extensive, and it is one that has consequences which reach far beyond the issue of transsexualism.

I would suggest that what we are witnessing here is a "benevolent form of behavior control and modification under the guise of therapy. It is not inconceivable that gender identity clinics, again in the name of therapy, could become potential centers of sex-role control for non-transsexuals — for example, children whose parents have strong ideas about the kind of masculine or feminine children they want their offspring to be, or women who stray too far from the limits of acceptable stereotypical behavior. But control here, as in the transsexual context, will enter and has already entered not with a bang but with a whimper. Conformity is enforced in the name of therapy, at the individual's request, and with the claim of 90 percent "satisfied customers."[7] If behaviorist theoreticians such as B. F. Skinner are right, and if behaviorist technicians such as Jose Delgado remain active, then the future social controllers can replace control through torture with control through pleasure, and supposedly with the individual's "informed consent."

In looking at transsexualism as a form of behavior control and modification, it is important to stress that as a proclaimed kind of therapeutic surgery, transsexualism lies on a historical continuum of similar medical ventures, all of which legitimate(d) body intervention for the purposes of improving behavior. In the nineteenth century, clitoridectomy for girls and women and, to a lesser extent, circumcision for boys were accepted modes of "treatment" for masturbation and other so-called sexual disorders. Clitoridectomy is still done today on millions of women, particularly in the northern parts of Africa. In the 1930's Egas Moniz, a Portuguese physician, received the Nobel prize for his "groundbreaking" work on lobotomies. Moniz operated on state mental hospital inmates, using this surgery for everything from depression to aggression. The new terminology for brain surgery of this nature today is "psychosurgery," which its proponents have attempted to disassociate from the earlier, "cruder" procedures used by Moniz et al. by pointing to its more "refined" techniques (e.g., electrode implants and stereotaxic brain surgery). Call it lobotomy or psychosurgery, surgeons, however, continue to intrude on human brains on the basis of tenuous localization theories which supposedly pinpoint the area of the brain where the "undesirable" behavior can be found and, moreover, excised. Finally, transsexual surgery is justified on the basis of adjusting a person's body to his or her mind.

What these surgical ventures have in common is that they derive their therapeutic legitimacy from a medical model which locates behavioral problems within certain affected organs. Surgery then alters (intrudes on), removes, and, in the case of transsexualism, adds organs. In each venture, a surgical fetishizing takes place in which the social and more expansive components of the issue are reduced to the most tangible and manageable locus.

Further, what each of these surgeries has in common is the modification and control of behavior. Clitoridectomies modify sexual behavior or fantasied sexual behavior; psychosurgery modifies the gamut of behavior from hyperactivity in preteen children to so-called manic depression in dissatisfied housewives; transsexual surgery modifies everything that comes under the heading of "masculine" or "feminine" in a patriarchal society — thus practically everything.

In the case of transsexualism, behavior modification is both a prerequisite for and an effect of the surgery. However, the controllers will emphasize that such surgery is sought voluntarily. They will point out that sex-conversion surgery is not forced on transsexuals. Like the "benevolent" behaviorism of B. F. Skinner in Walden II, transsexual surgery is presented as that which thousands seek, many of whom are turned away.

The ultimate effect of medicalized transsexualism is that a moral and political issue is transformed into a psychiatric and medical-technical problem to be solved by "passing" requirements, by hormone therapy, and by sex conversion surgery. Medical technicians focus on the surgical construction of desired genitalia. Artifacts of silicone breasts, artificial vaginas, and the like come to incarnate the essence of femaleness which the transsexual so desperately desires. The medical-technical solution thus begins to assert control in the narrow area of the chemical and surgical specialties, and once more life becomes amenable to medical values and medical solutions, and ultimately to the "triumph of the therapeutic." Since the result of transsexual surgery is that the transsexual becomes an agreeable participant in a society which encourages sexism, primarily by sex-role conformity, then ultimately the medical solution becomes a "social tranquilizer." What might have been social protest is stifled. The medical solution hides from the transsexual the possibility of being a history-bearing individual who, instead of conforming to sex roles, is in a unique position to turn his gender dysphoria into an effective protest against the very social structures and roles that spawned the dilemma to begin with. Thus the transsexual problem becomes manageable at the established level of sex-role conformity, and the psychological and medical clinicians become social engineers.

Transsexual surgery enables doctors to gain medical knowledge about the manipulation of human sexuality that probably could not be acquired by any other medical procedure. The actual procedures employed in sex conversion surgery serve to fashion bodies in conformity with feminine body stereotypes. Much of the medical literature is devoted to perfecting breast implants and functional vaginas in such a way as to bring the transsexual's body into line with a curvacious feminine figure.

What I have said thus far does not imply a callous or insensitive view of the desperate gender agony that persons who desire transsexual surgery must experience. It does, however, imply that sympathy should be put in the right place and that medical solutions which function under the guise of "sympathy for the oppressed," be exposed as iatrogenic (doctor-induced). Genuine sympathy should ask the *why* behind the suffering and propose ways of dealing with such suffering that will effect change on the deepest levels of the problem. Transsexualism is a half-truth insofar as it poses as an answer to the desperate experience of gender agony for many individuals in a role-defined society. But it is not a whole truth, in that it presents us with an inadequate solution which only reinforces the society and its socialization processes that produced transsexualism to begin with. To take a critical position on the issue of transsexualism is not to be less sympathetic or non-sympathetic to the real gender agony that transsexuals obviously endure, but rather to hope that all of those involved and concerned will be able to focus sympathy in the right place.

With these thoughts in mind, I would like to suggest several directions for change. On a philosophical level, transsexual treatment should be concerned about the integrity of both the individual and the society. Thus it would not encourage transsexuals to integrate their gender dissatisfaction into a "meaningful" conformity in which deeper questions of personal and social reality are not confronted. Instead it would help to supply the language needed to understand transsexualism within the social context of sex-role stereotyping and conformity. In other words, it would meet the problem on its own deeper, i.e., social ground, and not dislocate it to another, i.e., the medical-technical level. It would not replace gender suffering, which is very real suffering, with an artificially prolonged and synthetic maintenance of the problem so that the transsexual becomes an uncritical and dependent spectator of his most deeply decaying self.

More concretely, treatment should offer initially a type of counseling which incorporates elements of "consciousness-raising" about the deeper issues that lie behind the problem of *why* one finds oneself with a "female mind in a male body." I am not so naive as to think that this would make transsexualism disappear overnight, but it would at least begin to deal with some of the deeper issues that are involved. The experience of consciousness-raising groups within the women's movement has demonstrated that women can break the bonds of so-called "core" gender identity with the recognition that "the personal is indeed political."

Given peer encouragment to go beyond cultural definitions of both masculinity and femininity without changing one's body, persons considering transsexualism might not find it so necessary to resort to sex-conversion surgery. This may seem a lot to hope for, but it is no more than what has been achieved by "native-born" women who have taken seriously the ideals of feminism through the day-by-day living of a feminist awareness in a gender-defined society.

Thus, in the final analysis, I would ask transsexuals to join in the destruction of a role-defined society by not changing their bodies, but by instead proving both to themselves and to others that persons who experience gender dissatisfaction do not have to transsex in order to live fully in this society. Instead of becoming constructed women, transsexuals *as transsexuals* should take their unique form of gender dissatisfaction into their own hands and not be so willing to hand their bodies over to medicine. In doing so, they would also demystify the medical-technical solutions by regaining their autonomy and thus their own solutions.

Different perspectives on the issue of transsexualism need to receive more attention and publicity. We need to hear more from those women and men who at one time thought they might be transsexuals and who seriously considered surgery but decided differently — persons who successfully overcame their gender identity crises without resorting to the medically/technically controlled solution. We also need to hear more from professionals such as endocrinologist Charles Ihlenfeld who, after helping 100 or more persons to change their sex, left the field. Ihlenfeld decided that transsexual surgery only treats in a superficial manner, something that is much deeper.[8]

The issue, to my mind, is not a legal but a deeply moral one. As such,it requires that one ask different questions and find means to implement different answers. I am not advocating that sex conversion surgery be legally mandated out of existence. I am advocating that it be morally mandated out of existence. It is my contention that the elimination of transsexualism is not best achieved by repressive legislation forbidding such treatment, but by outlawing the conditions which make transsexualism possible (e.g., sex-role stereotyping in children's textbooks).

Transsexualism is merely one of the *most obvious* forms of gender typecasting in a patriarchal society. What may be overlooked is that these same stereotypes and behaviors are acted out every day in "native" bodies. The issues that transsexualism raises should by no means be confined to the transsexual context. Rather they should be confronted in the "normal" society that has spawned the problem of transsexualism.

NOTES AND REFERENCES

[1] I use the term, *medical model*, to mean an ideology that stresses: freedom from physical or mental pain or disease; the location of physical or mental disequilibrium within the individual or interpersonal context; an approach to human conflicts from a diagnostic and disease perspective to be solved by specialized technical and professional experts; intervention primarily by management and technical solutions.

[2] See especially, John Money and Anke Ehrhardt, *Man & Woman, Boy & Girl* (Baltimore: The Johns Hopkins University Press), 1972.

[3] Since the early 1960s, Money's work has been cited by feminist scholars — especially his theories about core gender identity being fixed at 18 months and his emphasis on the determining power of socialization. See, for example, Kate Millett, *Sexual Politics* (New York: Doubleday & Co., Inc.), 1970, pp.30-31. For further discussion, see Salzman's paper in this collection.

[4] John Money and Patricia Tucker, *Sexual Signatures: On Being a Man or a Woman* (Boston: Little, Brown, and Company), 1975, p.119. This book is merely a popularized version of *Man & Woman, Boy & Girl*.

[5] See especially, Robert Stoller, *Sex and Gender: The Development of Masculinity and Femininity* (New York: Science House), 1968, and *The Transsexual Experiment* (London: The Hogarth Press), 1975.

[6] Throughout this essay, I have consistently described the situation of the male-to-constructed female transsexual. I do this for several reasons. First, I intend to reinforce the fact that the majority of transsexuals are men. The accepted ratio is 1:4. Second, transsexualism is originated, supported, institutionalized, and perpetuated primarily by men. Third, although there are female-to-constructed male transsexuals, they are very real *tokens*. They are used to promote the universalist argument that transsexualism is a supposed "human" problem not uniquely restricted to men. The medical empire assimilates female-to-constructed male transsexuals to promote the illusion of the inclusion of women, and to claim that transsexualism is "sex-blind." For these reasons, I feel that to discuss *male and female* transsexuals in the same breath, or to use the words *he and she* when speaking of transsexuals, is to encourage the deception that transsexualism is a "human" problem equally affecting both men and women. For a more extensive discussion of this point and other ideas contained in this essay, see my forthcoming book: *The Transsexual Empire: The Making of the She-Male* (Boston: Beacon Press), 1979.

[7] Percentages vary slightly but most doctors and therapists who do postoperative follow-up report that "the majority" or "most of" the transsexuals they surveyed are satisfied, both with the results of the surgery and their own state of being after the operation.

[8] Quoted in "A Doctor Tells Why He'll No Longer Treat Transsexuals," *The National Observer*, October 16, 1976, p.14.

CONCLUSIONS

. . . Said the sages, "In the first place,
    The thing cannot be done!
And, second, if it *could* be,
    It would not be any fun!
And, third, and most conclusive,
    And admitting no reply,
You would have to change your nature!
    We should like to see you try!"
They chuckled then triumphantly,
    These lean and hairy shapes,
For these things passed as arguments
    With the Anthropoidal Apes.

. . . Cried all, "Before such things can come,
    You idiotic child,
You must alter Human Nature!"
    And they all sat back and smiled.
Thought they, "An answer to that last
    It will be hard to find!"
It was a clinching argument
    To the Neolithic Mind!

From the poem, *Similar Cases*, by Charlotte Perkins Gilman
Reprinted in the Foreword by Zona Gale, to the Harper and Row Edition (1963) of
The Living of Charlotte Perkins Gilman: An Autobiography, 1935.

The articles in this book point out many methodological flaws that underlie the current theories of biological determinism. Most of the difficulties derive from the fact that true objectivity is not possible for human beings rooted in cultural traditions.

In our Introduction we discussed the effects of the cultural context on the manner in which science is done. The questions we ask, the facts we notice and deem relevant, the theories that are the basis of further observations all are products of the society and culture in which scientists live, work and think. Furthermore, living in a society in which sex roles are important social categories, scientists' conviction that some sex differences in behavior are innate, as we have seen, has been a powerful force in directing a large body of work from the nineteenth century to the present. A society and a scientific mode of thought in which difference is more important than similarity has led to a bias within behavioral sciences toward seeing polarity and dichotomy, as Star points out in her discussion of the meanings of "male" and "female."

This book offers strong criticisms of the theoretical framework within which sex differences research is conducted and of some of the specific research methodologies as well as of the ways in which research results are interpreted and used.

Criticisms of Theories and Methods

The questions asked: The most fundamental problem is with the basic questions being asked. Science does not tell one what questions to ask; they are culturally determined. Science only offers methods for attempting to answer questions, and in so doing it puts some restrictions by focusing attention on the kinds of questions that are amenable to scientific resolution. In the case of the nature-nurture issue, from the point of view of science, the questions are incorrectly framed. One cannot determine scientifically how much of behavior is due to innate factors and how much to environment. The world cannot be dichotomized into organism and environment, for there are no organisms without an environment, and we cannot know an environment devoid of organisms. Similarly, the notion that we can separate and determine genetic and environmental components of social behavior is a conceptual fallacy, and one that cannot be overcome by inventing better technical and statistical tools.

Even the more restricted question about the origin of *variations* in behavior has severe limitations, for it only applies to variations within a given population in a given environment. No method exists at present by which results obtained in one environment or with one population can be used to give information about a different population or a different environment. There is no theory at present that enables one to determine the origins of behavioral differences between two groups in two different environments. This theoretical limitation invalidates all attempts to distinguish between genetic and environmental sources of sex differences, race differences or of any other *group* differences in behavior.

Limitations of the models used: Implied in the various papers is the question, whether the scientific method is applicable to the social sciences. In the physical sciences, models or theories are proposed and tested by means of experiments in which all the appropriate variables are (in principle) under the control of the experimenter. This type of procedure is not possible in the social and behavioral sciences since there is no way to hold variables constant. As Star points out, in such situations a controlled experiment is a figment of the imagination. Recognizing this, social scientists usually try to deal with the problem by using statistical samples, in order to establish the *range* of effects of the important variables. However, since there are usually a great many more important variables involved in anything that interests social or behavioral scientists than there are in the comparatively simple models required by physical scientists, statistics offers only a partial answer to the dilemma. Variables must be identified before their effects can be examined. Yet in general only a few variables are ever identified in the case of social science models, and little can be done to ascertain whether all the important ones have been taken into account.

Even if a model can be constructed that encompasses all the important variables, the social sciences, because they must deal with statistical samples, still can tell us nothing about individuals. For example, Alice Rossi argues that *on the average* men are not as good at parenting as women and that therefore they will need "remedial education." Yet this assumes that *all* or *most* men need this, although many men are acknowledged to be as good or better at parenting than the *"average* woman;" in the same vein, *all* women are judged not to need remedial training, though some are less apt than the male average. This is the kind of ridiculous situation that arises regularly when one tries to deal with individuals on the basis of (real or assumed) group averages.

It is an open question whether any of the current work in the social sciences pays attention to *all* the **important** variables that influence a given behavioral trait. Thus the correspondence of any model of human behavior with the real world is problematic. Maccoby and Jacklin's book, *The Psychology of Sex Differences*, discussed by Salzman, provides an example. The authors examine many studies of sex differences and conclude that there are relatively few differences and that they are subtle and difficult to measure. After reading the book, someone in another culture might well have the impression that in our society the behavior of males and females is virtually indistinguishable. Yet anyone who observes us, immediately notices obvious differences in the ways women and men act. Indeed Raymond's paper makes it clear that sex roles are so rigidly enforced that a male who wants to behave like a female can come to feel that he must change his *body* in order to do so and be socially acceptable.

Social science models have not proved to be very adequate in terms of reality tests and do not correspond well with how people behave in complex social situations. In this sense they are simply not very good science. Yet Western culture is so completely dominated by the modes of thought generated by the physical sciences that the conceptual framework they offer has been the only one available for the study of human behavior.

The use of animal models for human behavior: Given the difficulties of doing controlled behavioral experiments with people, animals have been widely studied with the idea that some valid analogies can be drawn from them. Our paper and Bleier's point out some of the difficulties of using animal models to distinguish between the effects of nature and nurture. First there is a difficulty in deciding what animals and which of the many things they do to take as models. Animals often behave in very diverse ways even within one species, their actions reflecting the interplay of different genetic characteristics with different environments. Related to this is the difficulty of determining whether a given animal behavior is "innate." There is increasing evidence that environmental differences can cause significant variations in the ways animals act and it is difficult to assess the extent of such effects since the interactions of organisms and their environments are complex.

Perhaps most hazardous is the assumption that, if innate, animal behavior can teach us something about the origins of "similar" human behaviors. There are several reasons for this. The very judgment that the behaviors are similar is questionable, since it involves the abstraction of behavior from its human or animal context and its treatment as though it were a thing in and of itself. (It also often involves the manipulation of language and of categories to imply that behaviors are similar when they may be significantly different, as we discuss later.) Further, there is no reason to suppose that similar behaviors in two species represent the same evolutionary solution to a given problem. The assumption that when animals from different species act in similar ways, their behaviors must share a common evolutionary and genetic origin ignores an important rule — the need to take species characteristics and environments into account.

The most important characteristic of people as a species is the development of our cerebral cortex in such a way that learning and culture have become our hallmarks. Evolution has given us a brain whose plasticity and versatility are its predominant features. Furthermore, as people we are involved in much more complex interactions with our environments than are other animals, since we ourselves create large portions of our surroundings. No animals interact with their environments in ways that are even remotely similar to the complex cultural interactions of people. Animal models involve entirely different environmental and behavioral contexts from the human ones, and therefore they are completely inadequate to test theories of human behavior, which is unavoidably interlocked with culture.

Selective and incorrect interpretations of data: Once a hypothesis has come to be accepted as close to certain, there is a strong bias toward noticing and accepting "facts" and "data" that support it. They become part of the foreground and are seen as important and noteworthy. (The same goes for ideas and interpretations that fit the cultural framework and that readily can be used to reinforce it.) Inconclusive data often can be ignored or, if this is not possible, interpreted in whatever way is most favorable to the theory. The work of Money and collaborators on sex hormones and the brain, discussed in several of the papers, is an example of the use of inconclusive data to reach "conclusions" that support cultural preconceptions.

As long as belief in the "paradigm" still holds, data which contradict it have several possible fates. They may be misinterpreted in a way that fits them into the paradigm. At times this can go so far as interpreting them to mean something other than what they do. Alternatively, things may actually be *seen* in such a way that they support the theory, as in the example of the playing cards we gave in the Introduction. Or finally, data can simply be ignored or judged to be invalid "artifacts." Such selective use of data — usually unintended — is obvious throughout the work on sex differences. The research discussed in this volume provides numerous examples, with sociobiology as a particularly rich source. Leibowitz gives a clear example of data being forced into a cultural framework or paradigm in her analysis of sociobiologists' treatment of hierarchy and dominance in primates. Given a basic view in which it is not envisioned that some primates have no hierarchies, their absence is described as "odd" and hierarchies are said to exist even when the evidence points to the contrary. Another example, discussed by Bleier, is the way in which evidences of women's contributions to human social and biological evolution are ignored in the "Man the Hunter" paradigm.

It is obvious that when scientists think that they have proved a theory, they must believe that they have the data to prove it. Since the questions biological determinists ask are flawed scientifically and the answers cannot be proved at this point, it is not surprising that there are many problems regarding the validity of the data.

Inappropriate use of language: As Salzman pointed out, in order to study a thing you have to define it. But behavior is complex and multidimensional, and hence not amenable to easy definition. Nor does it usually have a unique identity: behavior is not a thing and should not be reified and treated as though it were one. When two seals are rolling over and over together in the water, are they fighting or playing? Is it legitimate to transfer words from human cultural contexts to animals? Indeed, should words that carry considerable cultural connotations be used to describe animal behavior? Even with people, behavior often does not fall into easily definable categories. When two small children push each other, are they playing or fighting? Often it depends on the situation whether pushing is closer to fighting or to playing. Social gestures usually do not have a unique meaning and significance, but are defined by many and often subtle "undertones."

The behavioral sciences are particularly beset by the difficulties of linguistic ambiguity, difficulties that can lead to manipulation of definitions and descriptions to fit preconceived ideas and results. Vague and inconsistent use of language often can be used to assimilate unclear or even contradictory data. For example, the same word may be used for different behaviors, so implying that they are in some sense the same. The identity of the verbal label can then be used as *data* to support the notion of universals in human (and/or animal) behavior, when the behaviors in reality contradict it. Words such as aggression and dominance have been used in this way, as Bleier, Salzman and Leibowitz pointed out.

Another instance in which language has been used (at best) carelessly is when words which denote complex and highly variable economic and social institutions, such as "family" or "harem," are used to describe animal groupings or interactions. This practice reifies relationships and implies that they have an independent existence outside their specific human cultural contexts. Furthermore, giving the same name to human and nonhuman groupings by implication *asserts* their identity, which presumably is what one is trying to *prove*. In a circular way the identity of linguistic labels then "justifies" the usually unwarranted use of animal models for human behaviors.

The confusion of correlation with causation: In the behavioral sciences, as Star has pointed out, there is a tendency to assume that anything that is found to have a biological correlate must be innate. However, this assumption neglects the possibility that a given social environment itself may have important biological effects. Not only do we not know all the variables that are significant for any specific social behavior, but usually the relationships of the variables to each other are not known either. One does not know which are the dependent and which the independent variables. (Indeed, even in physics this choice is more contextual than "real.") Thus if we observe that a given behavior always goes with a particular physical trait, not only do we not know which causes which, we do not even know whether either of them causes the other or whether some third (perhaps undetermined) variable determines both. The assumption in studies of brain lateralization that correlation of left or right hemispheric activity with different cognitive abilities proves anything about their causation is a classic example of this confusion.

Effects of the Theories

Raymond focuses on the social and medical consequences of deterministic theories, and some of the other authors allude to them as well.

Such theories are needlessly limiting. If people think that they cannot do something, they probably will not try, since aspirations are usually kept within the limits of what is believed possible. In fact, one measure of "sanity" is that one not try to do something that is clearly impossible. For an ordinary person to try to fly or to be Napoleon is obviously insane; but it is also held to be less than sane to be afraid to try to do the things that *are* possible. Thus someone who is afraid to go out of the house is seen as neurotic at best. However, whether or not sane people try something depends not on what really *is* possible, but on what they and the people around them *believe* to be possible. Because theories of biological determinism give a false view of reality, they can lead to self-limitations that are not in accord with the real world. They also lead to the strengthening of social limitations by supplying rationales for the existence of social institutions that assign different individuals to unequal and confining positions in society.

Another deleterious effect of such theories is that they encourage the use of medical and technological methods of social control. Transsexual surgery for gender identity problems, brain surgery for homosexuals, lobotomies to "cure" aggression, all are means of dealing with people who deviate from social norms. Such pseudo-biological "solutions" are fostered by a framework of belief that sees biology as the controlling factor for behavior, which it is not. (But of course neither can complex behaviors be conditioned by *fiat*, except in very limited and defined situations.)

Responses to the Work

After looking at all the problems inherent in much of the work on the origins of sex differences in behavior, we must ask what responses we should make to it. What, other than calling attention to the scientific flaws and political implications of such research, ought one to do?

There is what we see as an unfortunate tendency on the part of some feminists to respond to these theories by inverting the arguments: to assert that biological differences in behavior between females and males indeed exist, but that women's behaviors and characteristics are superior to men's. Some make open avowals of this position, as for example, gina, as quoted by Star. Others such as Alice Rossi do this more indirectly. In her *Daedalus* article Rossi expresses the concern that by accepting male norms, feminists are trying to make women's behavior into that characteristic of white, middle-class men.

Her concern is well founded and important, but her response is to fall into the trap of positing women's special biological nature. A value judgment that one set of traits is more desirable than another is legitimate — we make such judgments all the time. But such arguments should not masquerade as "proofs" of the *biological* inevitability of certain traits in one sex or the other.

As a positive response to all deterministic research, we should repudiate any work that fosters stereotypes. Instead we must try to develop and encourage research that tries to understand how to develop complete human beings. We need to understand more about the different capacities of different people in different environments, rather than about the averaged, abstracted capacities of groups of people in some imaginary, homogenized reconstruction of contemporary Western societies. We need to encourage research into the complex ranges of interactions of people, cultures and environments in order to foster conditions that *expand* our possibilities rather than constrict them. Such work must look not only at effects of biology on behaviors, but at the effects (positive and negative) of behaviors and of environments on people's biology.

Marian Lowe
Ruth Hubbard

BIOGRAPHICAL SKETCHES

Ruth Bleier, M.D., from the Women's Medical College of Pennsylvania (now the Medical College of Pennsylvania), in 1949; experimental neuroanatomist; Professor in the Department of Neurophysiology and in the Women's Studies Program at the University of Wisconsin, Madison; practicing physician for several years; research on the hypothalamus; critical analyses of biological research on the brain and behavior and on women in the health care system; feminist political activist.

Ruth Hubbard, Ph.D., from Radcliffe College, in 1950; biologist; Professor in the Biology Department at Harvard University; research in the physiology and biochemistry of vision; present interests center on the politics of health and health care and on the sociology of knowledge and of science, particularly as they affect the lives of women; has written many articles and is co-editor of and contributor to a book entitled *Women Look at Biology Looking at Women; A Collection of Feminist Critiques,* to be published by G.K. Hall and Schenkman in Spring, 1979.

Lila Leibowitz, Ph.D., from Columbia University, in 1971; anthropologist; Associate Professor in the Department of Sociology/Anthropology at Northeastern University in Boston; research interests in biosocial evolution and social structures; author of a book entitled *Females, Males, Families: A Biosocial Approach,* published by Duxbury Press, 1978.

Marian Lowe, Ph.D., from the University of Minnesota, in 1963; chemist and environmentalist; Associate Professor in the Chemistry Department and member of the Women's Studies faculty at Boston University; past research interests in theoretical physical chemistry and solid state physics; present research on environmental issues, particularly energy policy and on the role of science in the creation of social ideology, particularly on the impact of science on women's lives.

Janice G. Raymond, Ph.D., Boston College, in 1977; Assistant Professor in Women's Studies/Medical Ethics at Hampshire College/University of Massachusetts, Amherst; research interests in genetic technology, and behavior control and modification; author of *The Transsexual Empire: The Making of the She-Male* (Boston: Beacon Press), 1979.

Freda Salzman, Ph.D., from the University of Illinois, in 1953; physicist; Professor in the Physics Department at the University of Massachusetts, Boston; research in the theory of particles and fields; member of the Sociobiology Study Group of the Boston Chapter of Science for the People; concerned with the uses of science in our society, has been most active in the area of effects of sexist science on women; has written articles on scientific sexism and has just completed an article on sociobiology.

Susan Leigh Star, doctoral candidate in Human Development and Aging, at the University of California, San Francisco; Poetry Editor for *Sinister Wisdom* and West Coast Editor of *Matrices: Lesbian Feminist Research Newsletter;* research interests in philosophy of science, feminism and aging.